24.95/18~1   C

Doing
Good Science
in Middle School:
A Practical Guide to
Inquiry-Based Instruction

DISCARD

Doing
Good Science
in Middle School:
A Practical Guide to
Inquiry-Based Instruction

By Olaf Jorgenson,
Jackie Cleveland,
and Rick Vanosdall

NATIONAL SCIENCE TEACHERS ASSOCIATION

Arlington, Virginia

NATIONAL SCIENCE TEACHERS ASSOCIATION

Claire Reinburg, Director
Judy Cusick, Senior Editor
Andrew Cocke, Associate Editor
Betty Smith, Associate Editor

ART AND DESIGN, Linda Olliver, Director
PRINTING AND PRODUCTION, Catherine Lorrain-Hale, Director
    Nguyet Tran, Assistant Production Manager
    Jack Parker, Electronic Prepress Technician

NEW PRODUCTS AND SERVICES, SciLinks, Tyson Brown, Director
    David Anderson, Database and Web Development Coordinator

NATIONAL SCIENCE TEACHERS ASSOCIATION
Gerald F. Wheeler, Executive Director
David Beacom, Publisher

Library of Congress Cataloging-in-Publication Data
Jorgenson, Olaf.
    Doing good science in middle school : a practical guide to inquiry-based instruction / by Olaf Jorgenson, Jackie
Cleveland, and Rick Vanosdall.— 1st ed.
        p. cm.
    Includes bibliographical references and index.
    ISBN 0-87355-232-6
    1. Science—Study and teaching (Middle school) 2. Inquiry-based learning. I. Cleveland, Jackie. II. Vanosdall, Rick.
III. Title.
    Q181.J69 2004
    507'.1'2—dc22
                                    2004012339

NSTA is committed to publishing material that promotes the best in inquiry-based science education. However, conditions of actual use may vary and the safety procedures and practices described in this book are intended to serve only as a guide. Additional precautionary measures may be required. NSTA and the authors do not warrant or represent that the procedures and practices in this book meet any safety code or standard of federal, state, or local regulations. NSTA and the authors disclaim any liability for personal injury or damage to property arising out of or relating to the use of this book, including any of the recommendations, instructions, or materials contained therein.

Permission is granted in advance for reproduction for purposes of classroom or workshop instruction. To request permission for other uses, send specific requests to: NSTA Press, 1840 Wilson Boulevard, Arlington, Virginia 22201-3000. Web site: *www.nsta.org*

SC*i*LINKS. *Featuring SciLinks —a way to connect text and the Internet. Up-to-the-minute online content, classroom ideas, and other materials are just a click away. Go to page xii to learn more about this educational resource.*

Contents

Chapter

1

The Demands of the Middle School Learner: Socialization, Autonomy, and Structure 1

Chapter

2

The Cornerstones of Good Science: Inquiry and Collaboration 9

Chapter

3

What Good Science Looks Like in the Classroom 17

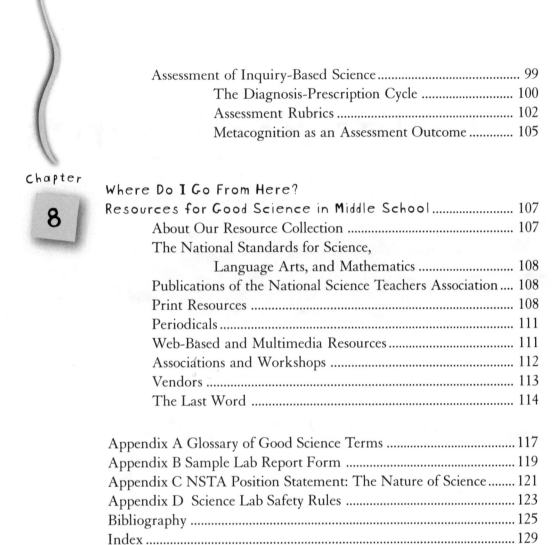

Chapter

8

Where Do I Go From Here?

Reviewers for This Book:

Vicki Baker
Science Teacher
National Board Certified Teacher
  in Early Adolescent Science
Alvarado Middle School
New Haven Unified School District
Union City, California

Janet Coffey
Assistant Professor
Curriculum and Instruction
College of Education
University of Maryland, College Park

Linda Froschauer
Science Department Chair and
  Science Teacher
Weston Middle School
Weston, Connecticut

Inez Fugate Liftig
Field Editor, *Science Scope*
Grade 8 Science Teacher
Fairfield Woods Middle School
Fairfield, Connecticut

Tables and Figures

Acknowledgments

The authors wish to thank the incredible people who inspired, reviewed, and/or helped craft this book. Our special gratitude goes out to reviewers/critics/therapists Lynn Dray, Eileen Gratkins, Janece Larson, Thea Hansen, Janey Kaufmann, Perry Montoya, Pat Dustman, Susan Sprague, and Mary Jane Strickland. We also appreciate Rock Leonard for his time and photography, and our families for their patient support.

DEDICATION

For Juliette, Wesley, Grant, and children everywhere
who will benefit from good science and great teachers

About the Authors

Olaf Jorgenson served as director of K–12 science, social sciences, and world languages in the Mesa Unified School District, Mesa, Arizona. Previously he was a teacher and administrator in U.S. and international schools, mostly at the middle and junior high levels. Ole is past president of the Association of Science Materials Centers (ASMC) and was on faculty with the National Science Resources Center's Leadership Assistance for Science Education Reform (LASER) strategic planning institute, with a focus on middle school science issues. He has presented on middle school science reform at ASMC's Next Steps Institute and is a past member of the National Science Teachers Association's National Committee for Science Supervision and Leadership. Ole's other publications focus on topics in school improvement, leadership, and science education. He holds a doctorate in educational leadership from Arizona State University. Ole lives in Kamuela, Hawaii, with his wife, Tanya, and their daughter, Juliette, where he is head of school at Hawaii Preparatory Academy. He can be reached at *ojorgenson@hpa.edu*.

Jackie Cleveland is a K–6 science specialist in Mesa, Arizona, where she has also served as a basic skills specialist. She is the recipient of the Presidential Award for Excellence in Science Teaching and a 1991 semifinalist for Arizona Teacher of the Year. She has served as the National Science Teachers Association preschool/elementary director, president of the Arizona Science Teachers Association, and advisory board member for CHEM at the Lawrence Hall of Science and participated in the development of the *Mr. Wizard's Teacher to Teacher* television production. She created, developed, and presented a professional development video series,

Science Scope. She teaches science methods courses at Arizona State University, Northern Arizona University, and University of Phoenix. Jackie lives in Mesa with her husband, Neal. Jackie's e-mail address is *jrclevel@mpsaz.org*.

Rick Vanosdall serves as associate director for research at Northern Arizona University's K–12 Center, a multi-university consortium dedicated to K–12 curriculum and instruction reform statewide (see *http://azk12.nau.edu*). Rick was previously the science specialist for Mesa, Arizona's, 22 junior and senior high schools, and prior to that worked as the science specialist for the district's National Science Foundation Local Systemic Change Grant for enhancing teacher performance in K–8 math, science, and technology. Rick is a 15-year high school science educator and taught chemistry, biology, ecology, and field biology in Mesa. Rick has conducted multiple national conference presentations for the Association of Science Materials Centers' Next Steps Institute and the National Science Teachers Association. Rick earned his doctorate in educational leadership at Arizona State University. He lives in Mesa with his wife, Kim, son, Grant, and daughter, Wesley, along with the family pooch, Copper. He can be reached via e-mail at *Rick.Vanosdall@NAU.EDU*.

Ray Turley, the illustrator, is a science teacher in Mesa, Arizona, with extensive experience at the junior high level. Ray's e-mail is *rgturley@mpsaz.org*.

Background

This book grew out of the authors' experiences while they worked in the Mesa Public Schools (MPS) in Mesa, Arizona. MPS is a metropolitan district with about 75,000 students and 90 schools,

K–12. Mesa has used learner-centered methods, inquiry principles, and hands-on science units since 1974, with a science resource center for kit development and distribution starting in 1979. Mesa's science program has earned praise from the National Science Teachers Association, *Harvard Educational Review, Newsweek, American School Board Journal, Parenting Magazine, American Scientist,* and *The Executive Educator.* In the past decade, the district has celebrated five awardees of the Presidential Award for Excellence in Science Teaching.

The district's science program and resource center was developed by longtime director Dr. Susan Sprague, now a science consultant living in semiretirement in northern Arizona. Mesa's resource center refurbishes and distributes over 10,000 kits annually to its 55 elementary schools. The middle school program also includes two self-contained fifth-grade Flight Centers with aircraft and helicopter simulators and night-vision goggle stations, serving all of the district's 4,000 or so fifth graders each year. To find out more, please visit the MPS Science and Social Sciences Resource Center (SSRC) Web site: *www.mpsaz.org/ssrc*

How can you and your students avoid searching hundreds of science Web sites to locate the best sources of information on a given topic? SciLinks, created and maintained by the National Science Teachers Association, has the answer.

In a SciLinked text, such as this one, you'll find a logo and topic near a concept your class is studying, a URL (*www.scilinks.org*), and a code. Simply go to the SciLinks Web site, type in the code, and receive an annotated listing of as many as 15 Web pages—all of which have gone through an extensive review process conducted by a team of science educators. SciLinks is your best source of pertinent, trustworthy Internet links on subjects from astronomy to zoology.

Need more information? Take a tour—*www.scilinks.org/tour*

Preface

A middle school science classroom was once described to us as "a nuclear reaction about to happen, on an hourly basis." At the time, that description was meant to illustrate the unstable, unpredictable, and at times irrational behavior of a mob of middle schoolers. Years later, we know that the behavior in question is pretty typical, but can be significantly more challenging to deal with when middle grades students are confined to neat rows of desks and numbed by textbooks, teacher-centered instruction, and lack of meaningful interaction with peers or their teachers. In this book we propose opportunities for learning and teaching amidst the sound and fury of a different sort of explosive (but productive) middle school science classroom. In our experience, good science—by which we mean inquiry-based science instruction—promotes the unexpected and delightful development of adolescent middle school students.

What is inquiry-based science instruction? It is "the creation of a classroom where students are engaged in (essentially) open-ended, student-centered, hands-on activities. This means that students must make at least some decisions about what they are doing and what their work means—thinking along the way" (Colburn 2003). As the National Science Education Standards say in a slightly different way: "Learning science is something students do, not something that is done to them" (NRC 1996, p. 2). Thus, good science or inquiry-based science is a shift away from textbook-centered, direct instruction that emphasizes discrete factual knowledge claims and passive observation of science phenomena, toward active, learner-centered, hands-on, minds-on investigations conducted to a greater or lesser degree by students themselves. (See Chapter 2 for a more thorough discussion of inquiry-based instruction.) Good science and middle school learners, we assert, are very compatible, as we'll explain later.

Who are we? We are three educators who have worked together in Mesa, Arizona, a school district that has embraced inquiry-based science instruction since 1974. We are among those who have come to enjoy the blossoming intellects, often comical behaviors, and insatiable curiosity of middle schoolers and who *choose* to work with them! With 55 years' combined experience in the profession, we've gathered a lot of ideas to share. We know from our interactions with educators around the country that precious few resources exist to assist science teachers "in the middle." A quick ERIC database search confirms this impression, and it was a central impetus for writing *Doing Good Science in Middle School*.

Our book is aligned to the National Science Education Standards (NSES), which set forth six areas defining what science teachers at all grade levels should be able to do:

- *Plan inquiry-based science programs*
- *Take actions to guide and facilitate student learning*
- *Assess teaching and student learning*
- *Develop environments that enable students to learn science*
- *Create communities of science learners*
- *Plan and develop the school science program* (NRC 1996, p. 4)

We've taken these and other NSES as our charge, using them as the basis for recommendations to assist new and experienced middle grades teachers. We define *middle school* as grades 5–8, consistent with the NSES, and throughout the book we keep self-contained team formats as well as departmentalized middle school configurations in mind.

Our work here is meant to meet other important objectives and to reach a variety of audiences, but above all, we intend it to be teacher-friendly. We wrote *Doing Good Science* as practitioners, for practitioners. In this book, you will find

- a comprehensive overview of inquiry-based middle school science instruction based on the NSES;
- information on best instructional practices and useful print and Web-based resources, science associations, workshops, and vendors;
- a conscious connection to the reading and writing skills that help determine— and are fostered by—student success in inquiry-based science instruction;
- 10 teacher-tested activities that integrate science with reading, writing, and mathematics (with an emphasis on relevant safety issues);
- a solid foothold for new teachers to help them teach inquiry-based science while better fathoming their often enigmatic middle grades students; and
- an opportunity for experienced inquiry-based teachers to reaffirm that what they do is "good science."

We hope readers will find this book easy to use. It can be read in its entirety or perused section by section as a reference for lesson and unit planning and as a basis for evaluating and modifying existing lessons. It will help teachers explain to their principals why their classes at times need to be noisy, bustling, and "social."

We understand that the general public might be skeptical about the reality of good science in the middle grades in light of the disappointing reports on science instruction appearing frequently in the media. Tests such as the annual National Assessment of Educational Progress (NAEP) (2003) and international comparisons such as the Third International Mathematics and Science Study (TIMSS) (1997) point to the middle grades as the place where science instruction begins to fall apart, after very strong performance by America's fourth graders on the TIMSS test (Schmidt et al. 1999).

However, we wrote this book as a *celebration* of effective middle school science, and not as a proposed cure for poor standardized-test performance. Indeed, in later chapters we look at innovative and engaging methods already in place in some schools for years—methods that, as recent research indicates, not only promote good science but also have contributed directly and dramatically to increased test scores in reading, writing, and mathematics, as well as science (Einstein Project 1999). Significantly, these schools serve some of our least advantaged student populations (e.g., Klentschy, Garrison, and Amaral 2001).

We hope this book is for some readers a point of departure from relying solely on passive text- and worksheet-dependent curricula and teacher-centered methods in favor of the active learning potential and rich teaching opportunities that inquiry-based science instruction makes possible in the middle grades. Teachers and students in middle schools from Anchorage, Alaska, to Fairfax County, Virginia, are already "doing" middle school inquiry science, and in select districts nationwide, some teachers are heading into a *fifth decade* of employing the principles and processes described in this book.

We'll look at the challenges, too. Some teachers shy away from inquiry because it seems daunting and a lot of work, but we've found that if teachers move slowly, inquiry activities are actually *less* labor intensive than traditional methods, once teachers establish a system and their students know the procedures. It is true that a shift to inquiry science can involve a substantial fiscal commitment for schools and districts that decide to invest in a science kit program. (Our activities, and philosophy, do not require kits, though they are definitely an asset.) Powerful national-level training is in place,

promoted by such organizations as the National Science Teachers Association, the National Science Foundation, the Smithsonian Institution, and the National Science Resources Center. Depending on the needs of teachers, schools, and systems, these trainings will take a district from strategic plan to science education revolution as quickly as the district's constituents can prepare for the necessary investment and overhaul. Strong preservice preparation, ongoing professional development, and improved approaches to teacher evaluation will ensure that good science is taught in middle school classrooms consistently and sustainably over time.

The most important question, though, is what the impact of all of this will be on youngsters in our schools. Is there evidence—that is, empirical data as well as sufficient testimonial and anecdotal reports—that inquiry methodology will significantly benefit middle-level children? That is: Is inquiry-based, standards-centered instruction good for middle school kids? Definitely! We'll show you why.

Let the journey begin!

References

Colburn, A. 2003. *The lingo of learning: 88 education terms every science teacher should know.* Arlington, VA: NSTA Press.

Einstein Project. 1999. *Cornerstone study.* Available online at *www.einsteinproject.org/studies/cornerstone*

Klentschy, M., L. Garrison, and O. Amaral. 2001. *Valle Imperial Project in Science (VIPS): Four-year comparison of student achievement data, 1995–1999.* Available online at *www.vcss.k12.ca.us/region8/Presentations.html*

National Assessment of Educational Progress (NAEP). 2003. Available online at *http://nces.ed.gov/nationsreportcard*

National Research Council (NRC). 1996. *National science education standards*. Washington, DC: National Academy Press.

Schmidt, W. H., C. C. McKnight, L. S. Cogan, P. M. Jakwerth, and R. T. Houang. 1999. *Facing the consequences: Using TIMSS for a closer look at U.S. mathematics and science education*. Boston: Kluwer.

Third International Mathematics and Science Study (TIMSS). 1997. Available online at *http://ustimss.msu.edu*

The Demands of the Middle School Learner

Socialization, Autonomy, and Structure

Whether you are a grizzled veteran of the middle-level trenches or a relative newcomer to science in the middle, your success with students in grades 5–8 will depend on your ability to adjust instruction to the cognitive, emotional, developmental, social, and psychological demands of the middle school learner.

Middle schoolers are "wired" for the active, inquiry-based approach to good science instruction. As we stated in the preface, by *good science* we mean a shift from textbook- and teacher-centered instruction to learner-centered, inquiry-based investigations in which the teacher becomes a guide and resource rather than a director and source. Inquiry, in turn, is an approach to scientific investigation that begins with a question, problem, or observation and is followed by testing hypotheses and reporting findings; it is the mechanism for varying degrees of student-directed discovery, autonomy, and heightened engagement in science activities.

Distinctions are often drawn between different *levels* of inquiry—for example, structured inquiry, guided inquiry, and open inquiry (Colburn 2003). In a structured inquiry activity, "the teacher gives students a (usually) hands-on problem they are to investigate, and the methods and materials to use for the investigation, but not expected outcomes. Students are to discover a relationship and generalize from data collected" (Colburn 2003, p. 20). In a guided inquiry activity, "the teacher gives students only the problem to investigate (and the materials to use for the investigation). Students must figure out how to answer the investigation's question and then generalize from the data collected" (Colburn 2003, p. 20). In an open inquiry activity, "students must figure out pretty much everything. They determine questions to investigate, procedures to address their questions, data to generate, and what the data mean" (Colburn 2003, p. 21). In this book, when we refer to inquiry-based instruction, we are referring to a combination of structured and guided inquiry. (A more thorough exploration of the meaning of inquiry-based instruction will be found in Chapter 2.)

The Match Between Middle Schoolers and Inquiry

In every key respect, good science addresses the fundamental needs of adolescent learners, especially when their teachers understand the profound developmental changes with which these youngsters contend. Conversely, traditional teaching methods that rely on textbooks, direct instruction, seatwork, and lecture are successful depending on students' ability to endure extended periods of concentration, inactivity, and careful note taking—none of which is a particular strength of most 10- to 14-year-olds we've known. However, such teacher- and text-centered instructional strategies prevail widely in the vast majority of hundreds of science classrooms we've observed.

Consider Table 1.1, illustrating the compatibility we've seen between the developmental traits of adolescents and the characteristics of inquiry-based instruction.

Inquiry-based teaching methods closely match the way middle-level students naturally learn, and support their intrinsic tendencies in developmentally appropriate ways. Middle school youngsters are not well-served by a curriculum founded on lecture and worksheets. Awareness of the developmental levels and cognitive parameters of 10- to 14-year-olds is paramount to effective middle school teaching in any subject.

TABLE 1.1

COMPATIBILITY BETWEEN MIDDLE SCHOOLERS AND INQUIRY

ADOLESCENT TRAITS	INQUIRY CHARACTERISTICS
Curiosity and interest in learning	Develops questioning skills; emphasis on "doing" science
Varying cognitive levels (shift from concrete to abstract reasoning)	Uses multiple process skills, cultivates different learning styles
Need for relevance, to connect learning to prior knowledge	Emphasizes experiment, experience, problem solving
Increased sense of independence	Emphasizes discovery learning
Need for social interdependence	Emphasizes collaboration and cooperative, project-based learning
Relatively short attention spans; easily bored, easily distracted	Uses activity-based instructional design; rote learning de-emphasized
Need for validation; insecurity, fear of failure, developing self-concept	Provides opportunities for noncompetitive, authentic assessments
Simultaneous need for autonomy and structure	Offers degrees of teacher influence/ guidance (i.e., from "structured inquiry" to "full inquiry")
Need to be acknowledged as an "adult"	Learner-centered, rather than teacher-centered

Typically Atypical

Until the mid- to late 1990s, relatively little research focused on middle-level learners compared with studies of older and younger children. Indeed, there is such a range of maturity in middle school students, it is difficult to find agreement about what these young people should even be called: Adolescents? Emerging adolescents? "In-between-agers?" (George and Alexander 1993, p. 2).

Whatever we call them, middle school students are not children, and they are not adults; yet they can display characteristics of both groups, and in dizzying succession, with little or no way to predict which will happen next. One day they want recess, and the next they want to drive your car. Needless to say, this makes life interesting for their teachers. Middle school expert and author Rick Wormeli observes that "young adolescents are moving through one of the most dynamic stages of development in their lives. As teachers, we might have to bush-

whack through the hormonal tendrils on a daily basis, but it's worth the effort to find the gold inside each child" (2001, p. 7).

Adolescents are coming to terms with rapidly increasing knowledge of themselves, their world, and their place in it. Nancie Atwell, a veteran middle school teacher, author, and one of the first educators to define and celebrate how learner-centered strategies work with middle schoolers, puts it this way: "Their sense of themselves, the world, and the relationship between the two is challenged every day by their own needs and the demands of new roles" (1991, p. 27). Some are already launching into puberty and toward sexual maturity and may find it difficult to relate with their peers who still enjoy playing with dolls and Legos. Preteens are excited and excitable, "interested in virtually everything but not in anything for very long" (Bybee et al. 1990, p. 4). Mood swings are dramatic and common; most middle schoolers are developing a more complex sense of humor; reactions are typically extreme—"When they like something they love it; when they dislike something they hate it" (Atwell 1991, p. 30).

For some youngsters, the middle school years are exciting and affirming, and for others, awkward and unsettling. Teachers at this level need to be especially sensitive, and vigilant, toward their students who manifest symptoms of depression or worse; as Atwell writes, "the thorns of adolescence are real and cause real pain" (1991, p. 26).

In any case, this sweeping range of social, physical, emotional, and intellectual maturation makes it difficult to generalize about "what works" in teaching middle school students, beyond the need to diagnose and monitor developmental progress throughout the school year. "Getting young adolescents to pay attention and learn is 80% of our battle in middle schools. The rest is pedagogy" (Wormeli 2001, p. 7). On a very basic level, teachers of middle schoolers should have a tolerance for ambiguity and a fundamental willingness to be flexible. (They also need, and middle schoolers really appreciate, a sense of humor.) To some extent, we believe teachers can learn to be spontaneous—to find comedy in inopportune displays of bodily functions, to lose precious class time to an evasive cockroach on the ceiling—but if a prospective middle grades teacher finds that her need to finish a chapter or get through a lesson overrides her willingness to tolerate the odd "accidental" belch or to spend five minutes calmly swatting an overhead roach amidst much fanfare, she might not be cut out to work with this breed of student. Effective middle school teachers enjoy interacting with students at this age.

If flexibility and spontaneity are assets in a middle school teaching assignment, it is equally important for us to turn our attention to the adolescent need for structure.

Need for Structure

Given the profound, rapid, and sometimes confusing array of changes middle schoolers experience between grades 5 and 8, it is not surprising that they seek stability to varying extents and in different ways. Above all, middle-level students need the structure provided by classroom procedures because it makes school a safe, protected, predictable environment in a surrounding world that for many of them is as unstable and unpredictable as their own mood swings. Their newfound freedom is alternately thrilling and frightening, and their limited experience with autonomy leaves most adolescents in need of some degree of order and security. It is usual for adolescents to fluctuate between demanding independence and welcoming direction from adults who they know care about them as individuals.

A substantial amount of our foundational moral development takes place during these years, when young people begin to associate actions with consequences—that is, moving beyond "nothing is wrong until they are caught" (Wong and Wong 1998, p. 151)—and with the broader ethical structures that support moral judgments. What is right, what is wrong, and why? How far is too far, and why? Thus, middle grades students depend on and will necessarily test the structure provided by classroom expectations and rules. If a middle grades educator successfully teaches procedures, students adopt the routines as their own so that structure (and rules) become a source of comfort rather than confrontation.

If, on the other hand, the teacher posts the rules without taking time to teach them, points to them when a student gets out of hand, and demands explanations for "misbehavior," there's bound to be continued testing and experimenting from the students ("accidental" belches and tipped chairs are favorites, for example). Middle schoolers understand the difference between control and support, manipulation and respect; they resent being treated like children, even when they behave childishly. Mutual respect is part of their worldview—and nothing cultivates respect more successfully than teachers who frame their expectations and procedures in terms perceived by middle schoolers as "adult to adult." In our experience, respect is most likely to be enjoyed by teachers who understand and set their expectations with attention to the fears, insecurities, goals, demands, and motivations of middle grades youngsters. (See more on classroom management in Chapter 5.)

Middle School Thinker: Not an Oxymoron

Most people don't typically perceive middle school-aged youngsters as "thinkers," at least not in the academic sense. Middle schoolers are by nature eager to seek and discover—but how they pursue and what they do with information they encounter can vary dramatically from one student to the next. Adolescent learners are scattered on the continuum between what psychologist Jean Piaget termed *concrete* and *formal operations*. Students still functioning at the concrete level have difficulty understanding the relationship between variables, for example. Those who have progressed developmentally toward formal operations can make inferences and begin to reason deductively. At that stage, students can design controlled experiments and

> ## "Education is not the filling of a pail, but the lighting of a fire."
> —William Butler Yeats

determine relationships between multiple variables. Formal operational thinking leads to understanding increasingly complex modes of organizational schemata, such as the periodic table or the structure of DNA.

It is common to have a classroom with a few students who can reason only at concrete levels sitting next to students whose sophisticated, abstract thinking can make it interesting for the teacher to stay ahead from lesson to lesson. A one-size-fits-all, text-based approach to science does not readily accommodate the wide developmental differences between students in the middle.

We need to acknowledge that middle schoolers, regardless of their place on the cognitive continuum, ought to be treated as "thinkers" for two important reasons. First, the nature of inquiry methods allows for—indeed, demands—that teachers design activities with a range of abilities and aptitudes in mind. Whether the student's role is to accomplish limited, concrete tasks as a member of a group conducting a complex project or to be responsible for a synthesis of a group's findings involving assumptions and flaws in the group's experimental design, inquiry activities are equally appropriate. Our sample activities in Chapter 6 illustrate this point.

Second, middle schoolers should be treated as thinkers because no matter at what point the teacher finds students to be functioning cognitively, socially, and behaviorally in the first weeks of school, the teacher's task is to move them forward throughout the school year. We realize that this happens amidst much sound and fury during the middle years; noted brain-research consultant Pat Wolfe (1999) observes that middle schoolers are often "hijacked" by their emotional brains, making it extremely difficult for meaningful cognition to take place. There are many challenges to prolonged student attention in a middle school classroom, especially if accompanied by inactivity or inconsistent procedures. But again, as is the case with many aspects of effective middle grades instruction, regaining and maintaining student engagement mostly comes down to teaching clear procedures, reinforcing routines, and developing and maintaining relationships.

Passion for Discovery

Whether middle grades science students are stalled by high school level assignments that are still too abstract, or bored by elementary-caliber seatwork that is not at all challenging, we've observed that science is often made to seem hard, disinteresting, and/or irrelevant. Science in the middle grades needs to be fun, fundamental, and connected to the lives of adolescents. We've found that when we fail to meet their needs in this way, far too many youngsters in the middle grades are turned off to science.

When many students fail in the middle, whether in a certain class or in a sport or in their socialization, the impact is formative and at times life-altering. In many students, negative attitudes toward science are bred in the middle years, can fester in the face of multiple failures, and become progressively resistant to remedy in later, more difficult, and increasingly abstract science classes. Teachers in the middle grades are charged with lighting the fires of "finding out," cultivating the innate adolescent passion for discovery, rather than snuffing it out with too much lecture or too many worksheets.

Inquiry-based science calls for active teacher instruction. After the necessary planning and setup by the teacher, a key role for him or her is to move from group to group or lab station to station, helping facilitate investigations. Teaching good science is also intensive, no doubt about it, requiring more time for planning and preparation than traditional teaching. Good science demands patience, tenacity, passion, and the desire to grow as a teacher. But we've found that it requires comparatively much more effort to deal with boredom, behavior problems, and the *Ferris Bueller* effect ("Anyone? Can anyone tell me the answer? Anyone? Anyone?") that accompanied our early attempts to teach from a

textbook. Indeed, once you have your students hooked on good science and familiar with the procedures, they do most of the work, and the teacher gets to enjoy the hum (well, roar) of excited young people *doing science*.

At-Risk Students

No discussion of instructional approaches for the middle school learner would be complete without acknowledging those adolescents who collectively have been described as *at-risk*. This phrase has become a catchall in recent years and can be used to refer to students who are living in circumstances of poverty or amidst other disadvantages, engaged in substance abuse, subject to abusive home situations, working with learning disabilities, or contending with an attention deficit–related disorder—or who are, for whatever reason, less likely than their peers to make progress and eventually graduate. Among these and other setbacks to learning, poverty is the most pervasive common denominator among at-risk children, often reflected by squalid school facilities and insufficient instructional resources—though they might be taught by dedicated and talented teachers, working against great odds. As Jonathan Kozol (1991) writes in his epic study, *Savage Inequalities: Children in America's Schools*, "I often wondered why we would agree to let children go to school in places where no politician, school board president, or business CEO would dream of working" (p. 5).

The good news is that increasing evidence demonstrates that inquiry-based strategies can lessen the achievement gap for at-risk students, as well as for limited-English-proficient students (Von Secker 2001; Cole 1995). Because inquiry parallels the way we learn—actively, constructing as well as receiving and expressing knowledge, putting learning in a context created mostly by the learner—it reaches those students for whom traditional lecture and seatwork are especially ineffective in light of lack of preparedness, difficult home circumstances, and other hindrances to learning.

Overall, we recommend that teachers of at-risk middle school students acquire a vast repertoire of strategies aimed specifically at the needs and challenges of individual students. In essence, as our students become more and more diverse, so must our strategies for teaching them. Inquiry-based science instruction provides opportunities for all children, as we demonstrate later in this book.

References

Atwell, N. 1991. *In the middle: Writing, reading, and learning with adolescents.* Portsmouth, NH: Boynton/Cook.

Bybee, R. W., C. E. Buchwald, S. Crissman, et al. 1990. *Science and technology education for the middle years: Frameworks for curriculum and instruction.* Washington, DC: National Center for Improving Science Education.

Colburn, A. 2003. *The lingo of learning: 88 education terms every science teacher should know.* Arlington, VA: NSTA Press.

Cole, R. W., ed. 1995. *Educating everybody's children.* Alexandria, VA: Association for Supervision and Curriculum Development.

George, P. S., and W. M. Alexander. 1993. *The exemplary middle school.* 2nd ed. Philadelphia: Harcourt Brace.

Kozol, J. 1991. *Savage inequalities: Children in America's schools.* New York: HarperCollins.

Von Secker, C. 2001. Effects of inquiry-based teacher practices on science excellence and equity. *The Journal of Educational Research* 95(3):151–159.

Wolfe, P. 1999. Presentation to Mesa Public Schools, Mesa, AZ, April 16–17.

Wong, H. K., and R. T. Wong. 1998. *The first days of school: How to be an effective teacher.* Mountain View, CA: Harry K. Wong Publications.

Wormeli, R. 2001. *Meet me in the middle: Becoming an accomplished middle-level teacher.* Portland, ME: Stenhouse.

The Cornerstones of Good Science

Inquiry and Collaboration

The more one learns about inquiry-based instruction in light of the traits of middle school learners, the more evident it becomes that adolescents are ideal candidates for inquiry science. Middle schoolers need activity, and they get excited about solving actual problems in teams—collaborating—as opposed to routinely completing worksheets in the stifling isolation of their desks. We have briefly explored the nature of inquiry-based science instruction in the preface and in Chapter 1; here we examine it in greater depth with an emphasis on how it can be applied by teachers in classrooms and how *collaborative* inquiry activities can be productive with highly social, easily distracted middle schoolers.

The Nature of Inquiry-Based Science

For a look at the nature of inquiry in the context of good middle school science, we turn to the volume *National Science Education Standards* (NRC 1996) and its table entitled "Changing Emphases to Promote Inquiry" (see Table 2.1).

It is important that science as inquiry not be misinterpreted as something so prescriptive as "the scientific method." Inquiry involves acquiring and applying scientific knowledge, developing and using higher-order reasoning skills, and communicating scientific information and conclusions—but not by following any lockstep procedure. According to the NSES, science as inquiry can only be accomplished "when students frequently engage in active inquiries" (NRC 1996, p. 145). Further elements of inquiry are found in Figure 2.1.

The foundation for inquiry-based instruction is formed by the questions teachers ask to guide instruction, along with student questions and challenges to answers that come with practice in inquiry investigations. We look at how to use "focus" questions to guide inquiry investigations in Chapter 7.

Moving a teacher, school, or school system toward inquiry science at any level involves

SCILINKS.
THE WORLD'S A CLICK AWAY

Topic: Science as Inquiry
Go to: www.SciLinks.org
Code: DGS9

TABLE 2.1.
CHANGING EMPHASES TO PROMOTE INQUIRY

LESS EMPHASIS ON	MORE EMPHASIS ON
Activities that demonstrate and verify science content	Activities that investigate and analyze science questions
Investigations confined to one class period	Investigations over extended periods of time
Process skills out of context	Process skills in context
Emphasis on individual process skills such as observation or inference	Using multiple process skills—manipulation, cognitive, procedural
Getting an answer	Using evidence and strategies for developing or revising an explanation
Science as exploration and experiment	Science as argument and explanation
Providing answers to questions about science content	Communicating science explanations
Individuals and groups of students analyzing and synthesizing data without defending a conclusion	Groups of students often analyzing and synthesizing data after defending conclusions
Doing few investigations in order to leave time to cover large amounts of content	Doing more investigations in order to develop understanding, ability, values of inquiry and knowledge of science content
Concluding inquiries with the result of the experiment	Applying the results of experiments to scientific arguments and explanations
Management of materials and equipment	Management of ideas and information
Private communication of student ideas and conclusions to teacher	Public communication of student ideas and work to classmates

Source: National Research Council. 1996. *National science education standards.* Washington, DC: National Academy Press, p. 113.

rethinking priorities and principles. It also invites teachers to reconsider their role and relationships with students as "coinvestigators"—together sharing an investigation for which they do not necessarily know the outcome. (Even teachers may not be able to predict an experiment's re- sults, when students are charged to a greater or lesser degree with its design and implementa- tion.) Generally speaking, inquiry-based lessons involve a shift in teacher responsibilities to the students when compared with traditional science instruction (as illustrated in Table 2.2).

FIGURE 2.1
WHAT IS INQUIRY?

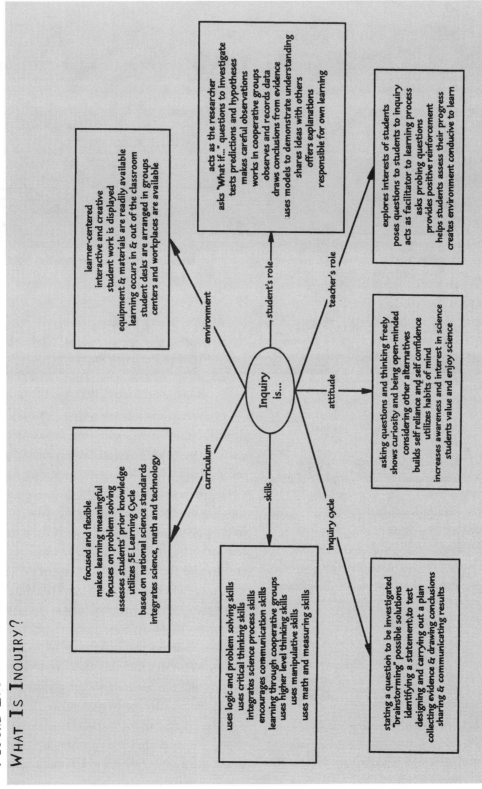

Source: D. Llewellyn. 2002. *Inquire within: Implementing inquiry-based science standards.* Thousand Oaks, CA: Corwin Press. Copyright © Corwin Press, Inc. Reprinted by permission of Corwin Press, Inc.

TABLE 2.2
INVITATION TO INQUIRY GRID

	DEMONSTRATIONS	ACTIVITY	TEACHER-INITIATED INQUIRY	STUDENT-INITIATED INQUIRY
POSING THE QUESTION	TEACHER	TEACHER	TEACHER	STUDENT
PLANNING THE PROCEDURE	TEACHER	TEACHER	STUDENT	STUDENT
FORMULATING THE RESULTS	TEACHER	STUDENT	STUDENT	STUDENT

Source: D. Llewellyn. 2002. *Inquire within: Implementing inquiry-based science standards.* Thousand Oaks, CA: Corwin Press. Copyright © Corwin Press, Inc. Reprinted by permission of Corwin Press, Inc.

The Teacher's Role in Inquiry

Inquiry demands a number of characteristic behaviors from teachers. Teachers need to

- *encourage and accept student autonomy and initiative*
- *use raw data and primary sources, along with manipulative, interactive, and physical materials*
- *use cognitive terminology such as* classify, analyze, *and* predict
- *allow student responses to drive lessons, shift instructional strategies, and alter content*
- *become familiar with students' understandings of concepts before sharing their own understandings of those concepts*
- *encourage students to engage in dialogue, both with the teacher and with one another*
- *encourage student inquiry by posing thoughtful, open-ended questions and asking students to question one another*
- *seek elaboration of students' initial responses*
- *engage students in experiences that pose contradictions to their initial hypotheses and then encourage discussion*

- *allow time after posing questions*
- *provide time for students to construct relationships and create metaphors*
- *nurture students' natural curiosity.* (Grennon Brooks and Brooks 1999, pp. 103–116)

In an inquiry science activity, the teacher directs without dominating and assists without answering. Asking reciprocal questions (what our students called "never giving a straight answer") drives students crazy at first; eventually they get used to it, and finally, they like it. But in the pursuit of professional science, no external authority provides hints to or reassures scientists, and the veracity of a scientist's conclusion, more often than not, is confirmed by her peers rather than someone outside the project design. In this way, inquiry-based science instruction mirrors the experience of "real" scientists. You can't fool the students. Once they develop the confidence and independence that "real science" affords, they know the difference.

The NSES call for teachers to develop students' skeptical abilities, encouraging them to be critical of appearances and assumptions much as the original Greek skeptics were. In this way, teachers model investigation and critical analysis, the heart of the process of "doing" science, and enable students to develop the necessary science process skills. (One teacher's slogan—plastered on every wall in his classroom, under the clock, on the trash cans—put the message this way to his students: *ASK QUESTIONS. QUESTION ANSWERS!*) This approach challenges teachers to let students make mistakes, as difficult as that may be, given the pressure teachers feel to cover content and complete the textbook by a set date.

That said, teachers using inquiry-based science experiences need to be sensitive to the perceptions of youngsters for whom active science may be entirely new. For some students in the middle grades, depending on their cognitive development and comfort with hands-on scientific procedures, inquiry activities can be very difficult. We've found that middle schoolers whose previous science success emerged from parroting back factual information are prone to initial frustration with inquiry investigation. Studies of best practice scientific methods demonstrate that a combination of teaching strategies, ranging from teacher- to student-centered, is most effective (Von Secker 2001) and is instrumental as teachers bring students from inquiry investigations to an understanding of scientific concepts. To put it bluntly: Sometimes kids just need the answers, and not the questions.

Strength in Numbers

Like the majority of contemporary scientific endeavors, inquiry activities are typically cooperative in nature. Granted, we know from experience that once in a while we all encounter a batch of kids who are simply *too* social to be productive in groups, even with excellent pro-

> "Nothing is certain. Even that."
> —Sextus Empiricus, the Skeptic (Greek philosopher, c. 160–210 A.D.)

cedures—and we understand that teachers have to be creative with these occasional "special crops." Overall though, good science at the middle school level incorporates many practices central to cooperative learning (as explained later in this chapter)—and it needs to, given the highly social nature of middle grades learners.

There are many excellent examples of cooperative strategies available to teachers in the professional literature, including several texts we list as resources in Chapter 8. Roger Johnson and David Johnson, well-known pioneers of cooperative learning in this country, list three essential principles of cooperative learning in *Cooperation in the Classroom* (Johnson, Johnson, and Holubec 1993):

- *positive interdependence–a "team" approach*
- *individual accountability*
- *face-to-face interactions* (p. 10)

Teachers can foster cooperative skills—listening and respecting others, encouraging, explaining, summarizing, checking for understanding, and disagreeing in a civil manner—by showing students that in their groups they must

- talk about the work,
- drill each other on the material,
- share answers,
- share materials, and
- encourage one another to learn.

One cooperative approach we find effective and readily adaptable to middle school classrooms is called "think-pair-share." Following an activity, demonstration, or discussion, students are asked to (1) think about what they just observed or talked about and (individually) write down two or three of the key ideas in their own words; (2) turn to their lab partners and listen to each other's ideas; and (3) move as a team to another group and repeat the sharing (or participate in a whole-group sharing). This strategy is both efficient and engaging.

"The brain," as Wormeli (2001) notes, "is innately social and requires interaction to remain elastic. In low-stimulus environments, a person's ability to think is diminished or never developed. In interaction-rich environments, thinking ability and memory increase" (p. 22). Of course, there's a difference between "stimulus-rich" middle school classrooms and simply "high-stimulus" classrooms. But as Wormeli discovered before brain research provided the explanation, "I *knew* those peer critiques, debates, and collaborative projects were doing something positive all those years. I just couldn't identify it" (2001, p. 22, author's emphasis).

Cooperative learning, however, is still perceived by some teachers as threatening because it requires them to relinquish some control depending on the lesson and activity. In inquiry, students are often up and moving, clustering in groups, comparing notes, negotiating, and generally making noise. For teachers to adapt to this classroom environment, they need a tolerance for ambiguity and a willingness to focus on student learning rather than on their own teaching as the primary force behind the lesson.

In our experience and in testimonials from middle-level teachers we interviewed for this book, we find that the number one reason that middle school science teachers shy away from collaborative inquiry activities is the fear of losing control. "Losing control looms as our greatest fear; rather than risk overstimulation we consciously choose not to stimulate" (Atwell 1991, p. 38).

Cooperative science activities in middle school need not be chaotic or anarchic at all (and we think middle grades principals will appreciate that!). Teachers using inquiry methods effectively—with focus questions, established procedures, and clearly defined objectives—are not "losing" structure, they are changing the *type* of structure. Middle schoolers need a degree of predictable order, and inquiry activities can be readily arranged to meet this need, provided they are used in tandem with other instructional methods suited to the differing demands of the middle-level learner.

In addition to "think-pair-share" and other cooperative strategies, inquiry activities conducted by students individually can also be a component of the middle-level science teacher's repertoire, as can inquiries based on a teacher demonstration of a lab activity. Compared with collaborative student inquiry exercises, these "shades of inquiry" may not capitalize on the adolescent's need for social interaction, nor will they necessarily enhance the student's ability to define an independent role in an interdependent team—as a scientist does. But they may be very helpful for students whose learning style isn't immediately compatible with an interactive group approach, who need to be coached to work more comfortably in a group pursuing an open-ended problem or question. The key is identifying and adjusting one's teaching methods to student needs through ongoing formative assessments, observations, and interaction and finding the methods that work for different groups of learners.

The point is, virtually any lesson can be adapted to involve inquiry, and inquiry lessons afford tremendous flexibility. Through the use of procedures taught, tested, and reinforced during the year, middle grades science teachers can lend structure to an activity by introducing a specific problem or topic, then allowing students the freedom to generate their own hypotheses and tests, to control the project design, and to work collaboratively toward project completion. This blend of structure and freedom is referred to as *structured inquiry* (as we discussed in Chapter 1, p. 1) and is an excellent match for the mix of order and spontaneity necessary to succeed with middle school learners. Structured inquiry complements effective cooperative learning, with its emphasis on individual accountability, interdependence, group interaction, and reciprocal feedback.

In our experience, inquiry "works" with students of all ages and ability levels because of the match between its methods and the way we learn. Learning lasts when new information is connected to previous knowledge and experiences, as constructivists hold. Especially for middle grades learners, the experiential platform built by inquiry activities provides relevance and context for the terminology, processes, concepts, and historical touchstones presented in text-based and teacher-centered instructional methods. Or, as noted cognitive scientist Larry Lowery puts it, "To learn geometry, we must have experience handling geometric forms and comparing them for similarities and differences. To learn about electricity, we must explore relationships among batteries, wires, and bulbs" (1992, p. 5).

On a practical note, we must stress that preparing for, setting up, and conducting inquiry activities (with or without kits) takes time. This is why we argue that it's not reasonable, much less instructionally advisable, to attempt open inquiry (see p. 1) *all the time*. We've found that an excellent guided-inquiry activity for teacher demonstrations is one in which the teacher asks the focus questions and students do the exploring, explaining, expanding, and evaluating; this approach is a convenient and an adequate alternative in a time crunch (and it also cuts down on safety risks with younger middle schoolers). We recommend that middle-level teachers present a blended array of demonstrations, short lectures, hands-on lab activities, and reflection exercises—in essence, a mix of teacher- and learner-centered instructional strategies. With this in mind, we move to Chapter 3, in which we present two teaching scenarios—from Mr. Hohum's class and Ms. Gottitrite's class—that illustrate the difference between traditional science teaching and inquiry-based teaching.

References

Atwell, N. 1991. *In the middle: Writing, reading, and learning with adolescents*. Portsmouth, NH: Boynton/Cook.

Grennon Brooks, J., and M. Brooks.1999. *In search of understanding: The case for constructivist classrooms,* rev. ed. Alexandria, VA: Association for Supervision and Curriculum Development.

Johnson, D. W., R. T. Johnson, and E. J. Holubec. 1993. *Cooperation in the classroom*. Edina, MN: Interaction Book Company.

Llewellyn, D. 2002. *Inquire within: Implementing inquiry-based science standards*. Thousand Oaks, CA: Corwin Press.

Lowery, L. 1992. *The biological basis of thinking and learning*. Berkeley, CA: University of California.

National Research Council (NRC). 1996. *National science education standards*. Washington, DC: National Academy Press.

Von Secker, C. 2001. Effects of inquiry-based teacher practices on science excellence and equity. *The Journal of Educational Research* 95(3):151–159.

Wormeli, R. 2001. *Meet me in the middle: Becoming an accomplished middle-level teacher.* Portland, ME: Stenhouse.

What Good Science Looks Like in the Classroom

Imagine you have been assigned to videotape two middle school science classrooms, located side by side, covering the same topic (Newton's first law of motion, let's say) on the same day with comparable student groups. You will enter each classroom silently, without disrupting instruction, and record what you see to compare the two encounters later.

In Mr. Hohum's room, here is what your camera records:

- It is dark except for the glare of the overhead projector.

- Students are sitting in straight rows, facing the teacher at the front of the room.

- Mr. Hohum is sitting next to his overhead machine, keeping one eye on the class while he writes lecture notes on the transparency.

- The classroom is quiet and orderly, and Mr. Hohum is the only one speaking.

- The teacher has defined Newton's first law and students are scribbling notes about objects at rest staying at rest unless acted upon by a force.

- The lights go on and Mr. Hohum directs the class to read a section in their text and complete a worksheet on Newton's first law.

This scenario illustrates a teacher-centered lesson, with Mr. Hohum "telling" and the students passively receiving information. They are learning about science, as opposed to "doing" science. If the principal were to drop in on this lesson, she might admire the orderly conduct of Mr. Hohum's students, their attention alternately fixed on the overhead, their note taking, and then textbooks and worksheets. Through exposure to this sort of teacher- and subject-centered routine on a daily basis, these middle schoolers have been sedated by inactivity and passive learning. Most of them probably aren't crazy about "science," either, thanks to Mr. Hohum's instructional methods. They have been programmed to "sit-get-spit-forget" whatever they're presented by the teacher, and, like Mr. Hohum himself, unfortunately they have counterparts in many middle school science classrooms across the country.

Stepping next door, your camera moves through the doorway of a very different learning environment in Ms. Gottitrite's class:

- The room is comparatively noisy, bustling with energy and simultaneous discussions at each separate station.

- Students are up and moving, working through procedures in small groups on two activities at different stations: a balloon car and a balls-and-ramps experiment designed to help students probe the relationship between force and motion.

- Some students at the stations are recording data and others are discussing their observations, even disagreeing on key points; some are testing, some are analyzing, some are preparing their reports.

- The teacher is moving between groups of students, asking questions, challenging them to expand and apply their observations to possible explanations of the phenomena they saw. She is carrying a clipboard to note her own observations about levels of student understanding—areas where she may need to reteach because her students didn't "get it."

In Ms. Gottitrite's class, the students are "doing" science, and the teacher is prompting them to connect their observations to possible explanations, to form hypotheses, experiment, gather data, and develop conclusions. This is learner-centered science, exemplifying the sort of classroom setting in which inquiry methods, problem solving, and other scientific practices flourish. Ms. Gottitrite's students are conducting experiments and building an experience base within which concepts like Newton's first law can take hold when the class shifts to a more directed study of the history of and theoretical basis for this essential principle of physics.

The 5E Method

This isn't to say there's no place for direct instruction methods, especially when a traditional lecture is needed to present challenging content.

Furthermore, some students are conditioned by lecture and note taking, and they struggle in their transition to inquiry-based instruction, even with teacher assistance. As a matter of well-rounded instruction, students should be exposed to some lecture and book work, and there are those learners who are best suited to direct instruction overall. Also, because middle schoolers are so social, some grouped table arrangements just aren't productive even with the best procedures, and teachers must adjust their methods (and table groups) accordingly. Our observations of effective science classrooms suggest that good science instruction involves a combination of approaches, including hands-on experiences in conjunction with direct instruction, independent research, and structured inquiry (for a discussion of *structured inquiry*, see p. 1).

Still, to accommodate different learning styles, instructional objectives, and developmental needs, the National Science Teachers Association recommends as a rule of thumb that 80% of time in middle school science classrooms be spent in active investigations (60% in the elementary grades and 40% in high school science) (NSTA 1990).

We recommend as a foundation the model of science instruction and lesson planning commonly referred to as the 5E method (Bybee 1997): *Engage, Explore, Explain, Elaborate, Evaluate.* These are the five basic procedures students will follow in pursuing inquiry-based science learning in the middle school classroom. In the guided-inquiry model that we find most effective with middle schoolers—detailed on page 1—the teacher determines the topic of inquiry and provides a discrepant event or focus question or problem to *engage* student interest and curiosity. The students, with their teacher as a guide and coinvestigator, begin to *explore* the problem or question. They make further observations and attempt to *explain* the phenomena they observe. The teacher then challenges students to *elaborate* on their understandings by linking observations to prior knowledge and by applying the concepts and skills in new situations. Finally, the teacher encourages students to *evaluate* their understandings and abilities, and the teacher *evaluates*, or assesses, the areas of strength and weakness exposed by student performance in the activity.

As part of the evaluation, the class reflects on where the inquiry can go next: What new questions do students have that can be investigated? In this way, inquiry fosters new inquiry. Important aspects of the 5E method are evident in Ms. Gottitrite's teaching vignette and in the activities in Chapter 6. The 5E method and student-centered strategies serve middle school science teachers and students by providing a rich context for

> ## "If teaching were the same as telling, we'd all be so smart we could hardly stand it."
> —Mark Twain

relevancy. According to master middle grades science teachers we interviewed for this book, such as Lynn Dray, a winner of a Presidential Award for Excellence in Science Teaching, one of the most important reasons to incorporate inquiry strategies into a broad repertoire of teaching tools is that such strategies help students make connections between conceptual science and experiential (real-life) science understandings. By "doing" science, students develop a set of experiences that give meaning to the scientific concepts and principles delivered in lectures and textbooks.

Learning Science as a Process

In the inquiry process, students acquire valuable skills through their interaction, collaboration, and problem solving with other students—skills that cannot be learned sitting in desks in rows and listening to a teacher. Learning science as a process, rather than a database of facts, is analogous to learning to play a musical instrument; we don't expect students to pick up piano by reading about it or watching the teacher play. These activities might be part of the overall strategy; but the students have to "do" piano in order to play it, just as middle schoolers must "do" science to master its concepts.

To look more closely at what we mean by good science, let's examine this definition of *science* agreed upon by a body of Nobel laureates (Klayman, Slocombe, Lehman, and Kaufman 1986):

> Science is devoted to formulating and testing naturalistic explanations for natural phenomena. It is a process for systematically collecting and recording data about the physical world, then categorizing and studying the collected data in an effort to infer the principles of nature that best explain the observed phenomena.

Note the key *action* verbs in this passage—science is *formulating, testing, collecting, recording, categorizing, studying, inferring, explaining*—all active, all dynamic, all procedural. (For a more detailed characterization of science teaching and learning, see Appendix C, "NSTA Position Statement: The Nature of Science").

"Learning science," as we quoted in the preface, "is something students do, not something that is done to them" (NRC 1996, p. 2). In this way, students become "coinvestigators" rather than vessels for information—they share with their teachers the responsibility for engaging in the process, the outcome, and the reflection during an investigation. This is part of a larger structural shift in our approach to science instruction in the middle grades (see Table 3.1).

Just as we urge that teachers and administrators collaborate to make these sorts of changes possible, we also believe in consulting the ultimate authority when it comes to determining what works and what doesn't: the students.

Student Response to Good Science

While traditionally educators haven't involved students in instructional decisions, we have observed over the past three decades that the reaction from students engaged in good science across the United States has been resoundingly enthusiastic. Quite simply, when students (and particularly middle school students) have an opportunity to engage actively, collaboratively, and frequently in inquiry-based instruction, science becomes their favorite subject.

Let us illustrate this student interest in inquiry-based instruction with a case example from our own district (the Mesa, Arizona, public schools). Inquiry-based science engages our young middle schoolers so effectively that several years ago, when science was inadvertently listed as "optional" on a seventh-grade preregistration form for junior high school, less than 5% of students at a feeder school then using traditional text-based approaches signed up for science—compared with over 95% of the sixth graders at a different feeder school with an inquiry-based program! Partly because of the overwhelming student enthusiasm generated by inquiry methods (and excellent sixth-grade teachers), inquiry-based instruction has become a firm component of our district's educational philosophy and process. The enthusiasm for science generated in the early grades translates to momentum for our students when they reach high school, reflected in the high class enrollments and pattern of student successes among our older students.

Those of you who are already "living" good science with middle schoolers have no doubt heard student comments such as these:

- (start of lab): "Oooh, this is gross!" (same student, end of lab): "I think I want to be a vet."
- "My brain hurts."
- "I hate you, Mrs. Nedergaard …. You make me think!"

TABLE 3.1

CHARACTERISTICS OF SUBJECT- VS. LEARNER-CENTERED INSTRUCTION

SUBJECT-CENTERED PROGRAM	LEARNER-CENTERED PROGRAM
"Told what to think"	"Explore how we think"
Learning a body of information	Learning how to learn
Mastery of concepts and skills	Creative exploration
Competition with other students	Cooperation with other students
Students follow teacher through lesson plans	Students explore with teacher guidance
Emphasis on teacher control of class activities and student actions	Student responsibility for learning
Teacher control	Student-teacher interdependence
Structured (six periods per day) schedule	Flexible scheduling
Whole group configurations	Variable group sizes
One teacher for a class	Varied structure, including team teaching
Textbook-dominated approach	Variable materials used in instruction
Outcome/answer predetermined by teacher	Presentation of results by students

Adapted from R. W. Bybee, C. E. Buchwald, S. Crissman, et al. 1990. *Science and technology education for the middle years: Frameworks for curriculum and instruction.* Andover, MA: National Center for Improving Science Education.

Good Science and the NSES

Inquiry-based science helps meet the goals set by the National Science Education Standards (NSES) (NRC 1996). The K–12 NSES were developed over the course of five years, following the 1989 publication of *Science for All Americans* by the American Association for the Advancement of Science (AAAS 1989) and its pioneering Project 2061. Project 2061 later produced the landmark *Benchmarks for Science Literacy* (1993), which defined scientific literacy for high school graduates; this work, along with the National Science Teachers Association's Scope, Sequence, and Coordination Project in the early 1990s, provided the impetus for the development of national science standards.

The NSES offer definitive direction for science teaching, professional development, assessment, content, and education system and program development—guidance that has given shape to science reform efforts nationwide. Many states have adapted their own standards to the NSES; and strategic reform initiatives such as the National Science Resources Center's companion programs, Leadership Assistance for Science Education Reform (LASER) and the Next Steps Institute (NSI), are modeled after the NSES teachings. Overall, the NSES content standards "present a broad and deep view of science teaching that is based on the conviction that scientific inquiry is at the heart of science and science learning" (NRC 1996, p. 15).

Note that the NSES consider inquiry to be a *content* standard, which for some teachers is a departure from the traditional model of "content" as limited to life, Earth, space, and physical science.

Principles of Scientific Knowing

The following principles of scientific knowing form the foundation for the inquiry-based approach to science education as advocated by the NSES and reinforced throughout this book:

- *The world is understandable.*
- *Scientific ideas are subject to change.*
- *Scientific knowledge is durable.*
- *Science cannot provide complete answers to all questions.*
- *Science demands evidence.*
- *Science is a blend of logic and imagination.*
- *Science explains and predicts, but is not authoritarian.*
- *Science is a complex social activity.*
- *There are generally accepted ethical principles in the conduct of science.* (AAAS 1990, pp. 2–11)

All of these characteristics of authentic science—as it is conducted by professional scientists and students of inquiry-based science alike—stand in contrast to much of what is presented to middle school students in science classrooms across the country. *We believe that the hurdles facing middle-level science reform involve improving methodology as much as enriching content knowledge and reforming preK-to-PhD science curricula.* Fundamentally, in the context of schooling, good science is inseparable from good teaching.

It's Good Teaching!

In their important review of research-based best practice instruction, *Classroom Instruction That Works,* Marzano, Pickering, and Pollock (2001) identify nine instructional approaches shown to improve student achievement across grade levels and subject areas. They bear a striking resemblance to the key elements of inquiry-based science instruction, as we know it. These approaches are

1. Providing guidance and practice in identifying similarities and differences
2. Providing guidance and practice in summarizing and note taking
3. Reinforcing effort and giving recognition
4. Including homework and opportunities for practice
5. Providing nonlinguistic representations (use of visual symbols, models)
6. Using cooperative learning strategies
7. Setting objectives and providing specific, corrective feedback
8. Generating and testing hypotheses
9. Incorporating cues, questions, and advance organizers

The authors identify these as best practices for all middle years content areas, not just science. Note also that these teaching behaviors are all ultimately matters of procedure, as are the exemplary strategies for lab safety, classroom management, collaborative projects, and other elements of inquiry-based science teaching in the middle grades presented in other chapters of this book.

In terms of down-to-earth advice about good teaching in science, we especially like the following guidelines from Lloyd Blaine, a science education consultant in Texas:

Good Science Basics

1. *Science smells. It also goes bump, boom, and makes other noises that make some teachers nervous.*

2. *Science is found everywhere, not just in textbooks.*

3. *Science is noisy and may appear out of control.*

4. *If you don't know the answer, tell your kids you don't.*

5. *Teach skepticism.*

6. *Bring science into other content classes and bring those other classes into your science class.*

7. *Teach students about scientists, both present and past, and how they did what they did.*

8. *Keep parents involved.*

9. *Don't be afraid to try new things.*

10. *Ask for training.*

11. *Don't bite off more than you can chew.* (Blaine 2001, p. 17–18)

Unfortunately, in more than 80% of America's K–12 schools, students learn science primarily by reading textbooks and passively observing teacher demonstrations (NSTA 2001). Sometimes, we realize, a school or district curriculum or a high-stakes test does truly dictate the nature of science instruction in a school; but in many cases, teachers with permission to adjust will not (or cannot) do so. And too often in middle schools, students are actually falling asleep in their textbooks.

The philosophical, conceptual, and procedural reforms we advocate for middle school science instruction, and in particular the shift toward more inquiry-centered activities, emerge from the changing emphases stated in the NSES (see Table 3.2).

Research Support

Until relatively recently, little empirical research has been conducted to establish the connection between inquiry-based science and improved student achievement, at least that measured empirically by standardized test scores. This is beginning to change. Since the mid-1990s, results from schools and districts across the country have carefully documented the favorable effects of inquiry science in longitudinal student achievement studies (e.g., Stohr-Hunt 1996, Wise 1996).

These findings were also borne out in data from a Wisconsin study presented at the National Science Teachers Association's 2001 National Convention. The study indicated that following three years of kit-based inquiry science instruction, K–8 student test scores improved from 55% of students scoring "proficient" or "advanced" on the state's standardized science achievement tests to 80% performing at that level. In the Einstein Project's Cornerstone Study (1999), also in Wisconsin, results at five "Einstein" schools (using inquiry methods and science kits) demonstrated that students scored an average of 77% on the state science test, compared with 64% among students at five control schools in similar socioeconomic contexts but without inquiry methods or kits. In the Cornerstone Study, 81% of the students who regularly engaged in inquiry activities attained mastery of science terminology beyond rote memorization, while only 20% in the control schools performed at that level.

In a study conducted in the El Centro, California, School District, four years of data collection from the Valle Imperial Project in Science (VIPS) show that a districtwide focus on inquiry science instruction had a significant impact on K–8 student achievement in science as well as in math, reading, and writing. In addition to making major gains in their standardized science test scores, El Centro students in sixth grade who had received inquiry science

TABLE 3.2

CHANGING EMPHASES CALLED FOR BY THE NATIONAL SCIENCE EDUCATION STANDARDS

LESS EMPHASIS ON	MORE EMPHASIS ON
Knowing scientific facts and information	Understanding scientific concepts and developing abilities of inquiry
Studying subject matter disciplines (physical, life, earth sciences) for their own sake	Learning subject matter disciplines in the context of inquiry, technology, science in personal and social perspectives, and history and nature of science
Separating science knowledge and process	Integrating all aspects of science content
Covering many science topics	Studying a few fundamental science concepts
Implementing inquiry as a set of procedures	Implementing inquiry as instructional strategies, abilities, and ideas to be learned

Source: National Research Council. 1996. *National science education standards.* Washington, DC: National Academy Press, p. 113.

instruction for the full four years of the study scored approximately 35% better in math and 28% better in reading, on average, than their classmates who had not been exposed to inquiry-centered, kit-based science instruction. On the district's writing proficiency exam, sixth graders who had not received inquiry science instruction scored an average of 23% on the test, while those who had been taught using the hands-on methods with science kits for the full four years of the program scored an average of 89% (Klentschy, Garrison, and Amaral 2001). In some ways, the success that El Centro students realized in literacy and math performance—through *science* instruction—was surprising and unexpected, although it follows El Centro Superintendent Michael Klentschy's belief that "the skills of reading and mathematics are strengthened when taught using engaging, high interest content" of inquiry science, including science notebooks (Klentschy, Garrison, and Amaral 2001).

Note that El Centro is a high-poverty, mixed ethnicity, traditionally low-performing district. Emerging studies are likely to show that while inquiry-based instruction benefits all students, it especially offers promise for closing the science achievement gap between underprivileged and more advantaged students.

Implementation Challenges

While we are convinced—and our experience is reinforced by teachers across the country—that inquiry-based science is ultimately easier, much more valuable and rewarding, and more cost-effective than traditional text-centered approaches, we would be remiss if we didn't acknowledge some of the challenges teachers can experience in implementing inquiry-based instruction.

For example, there never seems to be enough time to explore topics deeply and still cover the required curriculum. Also, making time for preparation, setup, and cleanup (until students are trained to expect and follow procedures) and for transforming traditional lessons to incorporate inquiry techniques and varied formative assessment techniques are other significant issues.

In addition, the purchase, care, refurbishment, and storage of materials required for inquiry-based science can pose logistical problems. Further, the inquiry approach can be expensive (although it doesn't have to be—after start-up costs, consumables are cheaper than workbooks and reams of copy paper for worksheets). Some of us who are accustomed to teacher-centered methods struggle with the shift to being a "coinvestigator"—in particular, the management challenges in training students to adjust their behavior when class activities switch from direct instruction to hand-on activities, and back again. Furthermore, helping students, colleagues, administrators, and parents to embrace inquiry-based science requires no small amount of strategy, foresight, and planning.

We mention these challenges not to dissuade or discourage readers, but to help you prepare in advance to move gradually, to adapt incrementally, and to focus on short-term, achievable goals—given your comfort level and experience. The transition to good science in the middle school involves subtle shifts rather than sweeping changes. Many teachers have reported to us that once students are in the inquiry-based science groove—anticipating procedures, cooperating in setup and cleanup, and feeling comfortable about working independently of the teacher's constant guidance—activities are actually less work and far more rewarding than traditional "sit-get-spit-forget" science. Here's a testimonial from Eileen Gratkins, one of the premier junior high science teachers now working in the Mesa, Arizona, school district:

For me, inquiry science was a lot of work at first, mostly in learning to let go of the control and learning to ask probing questions. But now, teaching inquiry is much less labor intensive than other types of lessons I do. The students do most of the work and I do less.

In terms of the perennial concern with funding inquiry-based science, years of budget data from our own district resource center in Mesa are proof that hands-on science is cheaper in the long haul than textbook-based programs.

Considering the challenges and the assets of good science, we understand that change is daunting to many of us who are struggling to meet daily objectives and teach well with multiple demands on our energy and time. That's why we advocate subtle shifts, which translate, practically, to developing and implementing a couple of new units each year. Trying to do too much too quickly is counterproductive for you and for your students.

References

American Association for the Advancement of Science (AAAS). 1989. *Science for all Americans*. New York: Oxford University Press.

American Association for the Advancement of Science (AAAS). 1993. *Benchmarks for science literacy*. New York: Oxford University Press.

Blaine, L. 2001. Science is elementary. *CESI Science* 34(2): 17–19.

Bybee, R. W. 1997. *Achieving scientific literacy: From purposes to practices*. Portsmouth, NH: Heinemann.

Bybee, R. W., C. E. Buchwald, S. Crissman, et al. 1990. *Science and technology education for the middle years: Frameworks for curriculum and instruction*. Andover, MA: National Center for Improving Science Education.

Einstein Project .1999. Cornerstone study. Available online at *www.einsteinproject.org/studies/cornerstone*

Klayman, R. A., W. B. Slocombe, J. S. Lehman, and B. S. Kaufman. 1986. *Amicus curiae brief of 72 Nobel laureates.* Available online at *Amicus Curiae of 72 Nobel Laureates*

Klentschy, M., L. Garrison, and O. Amaral. 2001. *Valle Imperial Project in Science (VIPS): Four-year comparison of student achievement data 1995-1999.* Available online at *www.vcss.k12.ca.us/region8/Presentations.html*

Marzano, R. J., D. J. Pickering, and J. E. Pollock. 2001. *Classroom instruction that works: Research-based strategies for increasing student achievement.* Alexandria, VA: Association for Supervision and Curriculum Development.

National Research Council (NRC). 1996. *National science education standards.* Washington D.C: National Academy Press.

National Science Teachers Association (NSTA). 1990. *An NSTA Position Statement: Laboratory Science.* Available online at *www.nsta.org/position statement&psid=16*

National Science Teachers Association (NSTA) National Convention. 2001. Presentations by the National Science Resources Center (NSRC) and the Einstein Project. St. Louis, MO, March 24.

Stohr-Hunt, P. 1996. An analysis of frequency of hands-on experience and science achievement. *Journal of Research in Science Teaching* 33:101–109.

Wise, K. 1996. Strategies for teaching science: What works? *The Clearing House* (July/Aug.): 337–338.

Integration Is Key

Science, Literacy, Math, and Technology

How do we integrate reading, writing, and math with science? Good science instruction is by nature cross-disciplinary, weaving literacy and numeracy with problem solving, discovery, and other higher-order thinking skills. The increasing fragmentation of academic subjects—such as the separation of science from math and literacy—plays a role in the disappointing performance of U.S. students in international achievement comparisons such as the Third International Mathematics and Science Study (TIMSS) (Schmidt, McKnight, and Raizen 1997).

In many ways, good science is at the crossroads of a curriculum. More than any other core academic subject, science routinely incorporates key goals in literacy and mathematical reasoning, in addition to the procedural approach and higher-order problem-solving skills that science cultivates.

Science and Literacy

While science and math seem a natural fit, those unfamiliar with the nature of scientific investigation and inquiry activities may not be aware of the close compatibility between science and literacy. In fact, science and literacy involve many reciprocal cognitive skills, as seen in Table 4.1.

Veteran science educators recognize the connections between science, reading, and writing. Not only are many conceptual skills transferable between literacy and science (e.g., predicting, identifying cause and effect, and using evidence); reading and writing are also integral to good science instruction through science notebooks, lab reports, research projects, group presentations, and other elements of instruction that reflect national and state standards for the language arts. Middle school science teachers can collaborate with their colleagues in other fields to identify reading, writing, speaking, and listening activities that can be conducted across the curriculum. As with classroom management expectations (see Chapter 5), our experience tells us that efforts to have an impact on student outcomes in middle school are most likely to succeed if they are implemented schoolwide. That is because middle school learn-

TABLE 4.1

RECIPROCITY BETWEEN LITERACY AND SCIENCE SKILLS

LITERACY	SCIENCE
Note details.	Observe and retain small details.
Compare and contrast.	Make notes about the way a variety of substances react (e.g., to heat).
Predict.	Hypothesize about what will happen next.
Work with sequences of events	Work with processes of logic and analysis.
Distinguish fact from opinion.	Use evidence to support claims.
Link cause and effect.	Study what causes things to react in a particular way.
Link words with precise meanings.	Develop operational definitions of a concept through experiences.
Make inferences.	Infer based on observation and evidence.
Draw conclusions.	Conclude by combining data from various sources.

Source: M. P. Klentschy and E. Molina-De La Torre. 2004. Students' science notebooks and the inquiry process. In *Crossing borders in literacy and science instruction: Perspectives on theory and practice*, ed. E. W. Saul, p. 342. Wilmington, DE, and Washington, DC: International Reading Association and NSTA Press.

ers have a developmental need for structure and will respond best to instructional approaches that they encounter in a variety of classes.

In schools where achievement test data can be used to identify performance patterns in individual students—for example, test results indicating which mode of writing (research, narrative, expository, persuasive, or creative) is a student's weak area—science teachers can often be part of a team effort to improve student performance by adapting assignments accordingly. Further, schools that adopt a template used consistently by teachers in every subject to teach the writing process can reinforce key skills such as presenting main ideas, organization, voice, word choice, sentence fluency, syntax, writing conventions, and effective presentation.

"Research" is an objective listed in the national standards for language arts, math, and science (NCTE/IRA 1996; NCTM 2000; NRC 1996). Too often, though, middle school research projects become individual competitions between students (and parents) rather than opportunities for cooperative learning, research skill development, and meaningful interaction. In her acclaimed study of young writers and the writing process, *Living Between the Lines*, Calkins (1991) recommends using whole-class research topics to maximize cooperative learning opportunities. In the whole-class approach, students work in collaborative teams to research, record, and present components of a *single* topic or theme, rather than working on many independent projects covering individualized topics. This approach is compatible with the social needs of the middle school learner (as discussed in Chapter 1) and enhances the teacher's efforts to develop middle school learning communities. Whole-class research also makes it easier for students to gather resources, something that is particularly useful in schools with limited media center collections. We recommend that research and writing projects in middle school science courses incorporate a writing process (i.e., prewriting, drafting, revising, editing, and publishing) and strategies that are consistent with the students' other subjects. Narrative and lab reports, research projects, essays, biographies, even creative writing assignments can be used to promote a school's literacy goals.

Science Lab Notebooks

Whenever students are involved in inquiry-based science, whether in a laboratory setting or while conducting field research, they should have the opportunity (and be expected) to record data as they progress through the process. Teachers can help students to develop this essential habit by providing them with commercially prepared lab worksheets or simply giving them blank sheets of paper inserted in binders. We've found that middle-level students generate more original thoughts and observations when they start with a blank page than when they fill out a structured worksheet, although teachers will need to provide some degree of introductory preparation, especially for younger students, about what sorts of thoughts and observations are to be recorded on the blank pages.

Science lab notebooks are becoming prevalent in middle school science across the country. Students use them to record lab observations, describe findings, list questions or problems, practice journaling skills, and engage in critical reflection. Teachers may incorporate prompts into the lab activity that bring students back to the journals to think, write, and reflect. Notebooks also give the teacher a means of ongoing formative assessment to the extent that instruction is adjusted according to the needs and deficiencies identified in the notebooks.

Science lab notebooks are different from the personal journals frequently used in language arts classes, in part because they are more structured. A typical lab book activity begins with a question to investigate. Students then add a prediction of outcome, observations made, procedures used, and a conclusion. (See Appendix B, "Sample Lab Report Form.") Like journals, though, lab notebook entries often involve reflection, and in the case of inquiry, we know that observations and investigations lead to more questions. Teachers encourage students to cultivate skepticism and critical analysis in their lab notebooks based on students' experiences with hands-on activities. That is why it's important for teachers to provide time for students to write in their lab notebooks after completing investigations. Well-organized writing takes time... like good science!

We've found that teachers shouldn't assume that students automatically understand the purpose of lab notebooks or what's expected when writing in the notebooks. Students need to know how to use the notebooks and why scientists use them, including the value they hold for scientists whose work can lead to important discoveries, closely guarded secret findings, and sometimes patented solutions to problems. Students will develop ever-stronger notebooking skills through practice, especially if the teacher regularly views the notebooks, makes comments, and assigns a formal grade for completion and accuracy. (If teachers choose to formally evaluate notebooks, students should not have to guess at what it takes to get a good grade. We recommend that initially the notebooking process should be modeled and monitored for students so the expectations are clear. A checklist for evaluating the notebooks is found in Figure 4.1.)

Lab notebooks are your portal to your budding scientists, and can indicate how much of your instruction and influence is getting through.

Another asset of lab notebooks is that they can be infinitely manipulated to involve varied writing proficiencies and to meet specific na-tional and state literacy standards. In their roles as tools for inquiry and for literacy integration, science lab notebooks have been successful vehicles for improving student writing (Klentschy and Molina-De La Torre 2004).

Personalizing Literacy in Science

There are numerous ways teachers can make "scientific" writing more stimulating and creative for students. For example, when address-

Figure 4.1

Science Lab Notebook Checklist

The following checklist can be used to assess how well a student is keeping records during science activities. You may consider giving students a copy of the checklist to enable them to monitor themselves through the process.

General Information
- ■ Question being investigated is stated.
- ■ All entries are dated.
- ■ Writing conventions (e.g., punctuation, capitalization) are correct.
- ■ Presentation is clear.

Observations
- ■ Description is very detailed.
- ■ Description is complete.

Illustrations
- ■ Drawings are accurate.
- ■ Drawings are labeled.
- ■ Drawings are in color.

Procedure
- ■ List materials used.
- ■ Sequence steps followed.

Communication of Data
- ■ Graphic/table is complete.
- ■ Graphic/table is labeled.
- ■ Graphic/table is mathematically correct.
- ■ Written portions are clear and complete.

Analysis and Conclusion
- ■ Analysis is clear and logical.
- ■ Explanation is complete.
- ■ New questions or investigations are proposed.

Line of Learning
- ■ New learning is shared.
- ■ New curiosities are shared.

ing ecologically sound ways to improve the environment, you might ask students to

- write a paper convincing a neighborhood curmudgeon to recycle,

- write a letter to a legislator urging support of a clean air (or water) bill,

- outline a plan for cleaning up a neighborhood vacant lot or park,

- create a recycling jingle for radio to raise the level of listeners' concern, or

- create a 30-second "drought awareness" spot for television.

And this is just the beginning. Rick Wormeli (2001), a leading expert on teaching in the middle grades, notes that "science has many natural uses for writing, from lab reports to poetry. The blend of personal discovery and science that we might see in *National Geographic* or *Discover* magazines is achievable in our middle school classrooms" (p. 132). Wormeli recommends the following writing activities as appropriate and exciting options for students in middle school science courses:

- *Write the life story of a scientist.*
- *Make a schedule.*
- *Make up a tongue twister.*
- *Write instructions (procedures).*
- *Write a consumer's guide.*
- *Write an origins myth.*
- *Create a calendar in which the picture for each month shows a particular aspect of a scientific topic.*
- *Write a science fiction story.*
- *Examine a common scientific misconception, how it is perpetuated, and what can be done to correct it.*
- *Explain why another student obtained certain lab results.*
- *Create a board game focusing on the basic steps of a science cycle or principle.*

- *Research and write a report about a scientific discovery that changed the world.* (p. 132)

Reading and Science

In addition to the many writing opportunities that inquiry-based science instruction makes possible, it also helps students to develop reading comprehension skills. Teachers can use the following questions from Thier and Daviss (2002) to help students learn to reflect on science writings and develop their reading and research strategies:

Reading Comprehension Prompts for Students

- *Predicting:*

  With a title like this, what is this reading probably about?

  What will happen next? (Turn to your partner and tell what might happen.)

- *Reflective questioning before reading:*

  Why am I reading this?

  Why does the author think I should read this?

  What do I expect to learn from reading this?

  How does this relate to my life?

  What do I already know about this topic?

- *Reflective questioning after reading:*

  What do I still not understand?

  What do I still want to know?

  What questions do I still have about this topic?

- *Paraphrasing or retelling:*

  What was the reading about?

  Can I explain to my partner or group, in my own words, the meaning of what I just read?

- *Summarizing:*

  Can I identify all the key concepts from the reading and write a summary using these concepts? (pp. 42–43)

Good science instruction also incorporates speaking and listening activities through presentations, projects, discussions, and reports, which are other areas of skill development called for in state and national language arts standards.

A number of documented and forthcoming studies support the role of science as an effective content vehicle combined with instruction in discrete reading skills. In particular, science has been used successfully to reach limited English proficient and disadvantaged readers and significantly improve student achievement on standardized tests (Klentschy and Molina-De La Torre 2004). Cole (1995) recommends several strategies shown to encourage active reading skills, such as reading aloud (teachers to students, student to students) and "meaning-driven" reading—that is, reading rooted in investigations and problem solving. These strategies, of course, are compatible with inquiry-based science instruction.

Science IS Mathematics

Like writing and reading, math is an essential part of good science instruction. While the two disciplines are widely perceived as inseparable, however, too many middle grades science teachers miss the opportunity to reinforce mathematical concepts, skills, or modeling in a way that will enhance student proficiency and achievement. Math concepts are present in scientific operations such as graphing, predicting, measuring, weighing, and collecting and analyzing data, but teachers must help students see the connections between the mathematical processes that are embedded in science activities and the mathematical principles students are learning in their math classes.

Regarding grades 5–8, the NSES call for "mathematics that students should use and understand" (NRC 1996, p. 219)—that is, math is or should be a natural extension of middle grades science content and inquiry teaching benchmarks.

Specifically, middle school students should do math in science activities that challenge them to

- *Represent situations verbally, numerically, graphically, geometrically, and symbolically*
- *Use estimations*
- *Identify and use functional relationships*
- *Develop and use tables, graphs, and rules to describe situations*
- *Use statistical methods to describe, analyze, evaluate, and make decisions*
- *Use geometry in solving problems*
- *Create experimental and theoretical models of situations involving probabilities.* (NRC 1996, p. 219)

Beyond basic computation, well-planned inquiry science activities typically call for students to engage in estimation, proportionality, and even basic algebraic and geometric concepts—topics identified as weaknesses in U.S. students' science performance on the TIMSS international comparison (Schmidt, McKnight, and Raizen 1997). Middle school science teachers should focus on developing activities that explicitly engage students in these mathematical operations, especially algebra and geometry operations. Please refer to the 10 activities in Chapter 6 to see how mathematical processes may be incorporated in good science activities.

Math and science are sometimes lumped together as "aptitudes" that some students have, and others don't. In math as well as science, the middle grades are where students begin to develop an identity as a learner—successful or unsuccessful—in these subjects. As in science, many of our colleagues involved in math coordination and training believe that mechanical, teacher-centered methods are one reason middle schoolers perceive themselves as "bad at math." We believe that the integration of math into good science instruction must adhere to the

SciLINKS.
THE WORLD'S A CLICK AWAY

Topic: Math and Science
Go to: *www.SciLinks.org*
Code: DGS32

same inquiry-based instructional principles that we discuss throughout this book.

Good Science Can Be Low Tech

Looking at the NSES, it is important to distinguish between the common use of the term *technology* and its meaning as it relates to doing good science. The NSES stress that technology should be used to meet a need or solve a problem, but this doesn't mean you need computers and expensive electronic probeware to conduct inquiry activities. Technology can simply be the means to a solution and can involve little more complexity than challenging students to design and build shockproof containers used in an egg drop or to recombine everyday materials to invent a better mousetrap—without a mousetrap. According to the NSES, the use of technology "should be readily accomplished by the students and should not involve lengthy learning of new physical skills or time-consuming preparation and assembly operations" (NRC 1996, pp. 161, 165). In light of this statement, we can conclude that the NSES discount the need perceived by some teachers and schools to purchase complex computer programs or specialized hardware for science instruction. Good science is *not* dependent on hardware and software. Good science can be low tech.

That's not to say that we're opposed to using scientific technology, if a school or district can afford it. Indeed, we've worked with fifth graders in scaled-down electronic flight simulators to learn the physics of flight and with eighth graders using night-vision goggle technology to study rock formations under desert starlight. We encourage teachers to pursue technology resources and training to use technology, as many investigations can be enhanced by technology. We also appreciate the advantages of using prepackaged science kits, particularly when they are part of a systemic plan for inquiry science and accompanied by training from the kit vendor or through the Association of Science Materials Centers (ASMC). However, the absence of science kits or computer networks does not prohibit teachers from implementing inquiry methods or from providing students with exciting learning opportunities.

References

Calkins, L. M. 1991. *Living between the lines.* Portsmouth, NH: Heinemann.

Cole, R. W., ed. 1995. *Educating everybody's children.* Alexandria, VA: Association for Supervision and Curriculum Development.

Klentschy, M. P., and E. Molina-De La Torre. 2004. Students' science notebooks and the inquiry process. In *Crossing borders in literacy and science instruction: Perspectives on theory and practice,* ed. E. W. Saul. Wilmington, DE, and Washington, DC: International Reading Association and NSTA Press.

National Council of Teachers of English (NCTE) and the International Reading Association (IRA). 1996. *Standards for the English language arts.* Urbana, IL, and Newark, DE: NCTE and IRA.

National Council of Teachers of Mathematics (NCTM). 2000. *Principles and standards for school mathematics.* Reston, VA: NCTM.

National Research Council (NRC). 1996. *National science education standards.* Washington, DC: National Academy Press.

Schmidt, W., C. McKnight, and S. Raizen. 1997. *A splintered vision: An investigation of U.S. science and mathematics education.* Boston: Kluwer.

Thier, M., and B. Daviss. 2002. *The new science literacy: Using language skills to help students learn science.* Portsmouth, NH: Heinemann.

Wormeli, R. 2001. *Meet me in the middle: Becoming an accomplished middle-level teacher.* Portland, ME: Stenhouse.

Classroom Management and Safety

Welcome to the challenge of making good science come to life in your classroom. In this chapter we look at how to get the classroom ready for inquiry-based lessons and how to prepare students so their activities are engaging, productive, and safe.

The essence of classroom management in the middle school is teaching students to expect and to follow *procedures*. If middle grades teachers accomplish that task early, and reinforce it throughout the term, they can expect significantly fewer surprises and hassles when it comes to student behavior, lab safety, and upkeep of equipment and materials.

The First Days with Students

> *What you do on the first days of school will determine your success or failure for the rest of the school year. You either win or lose your class on the first days of school.* (Wong and Wong 1998, p. 3)

Harry and Rosemary Wong's assertion above appears bold, but it is immediately relevant to teaching in the middle grades. Teachers preparing for the first days of middle school science need to tackle the following crucial issues before students ever enter the classroom door:

■ classroom arrangements that enable success (e.g., strategic groupings of students, posting classroom procedures and safety expectations, posting exemplary student work around the classroom),

■ an idea of the instructional "big picture"—that is, where you want to go, in keeping with the National Science Education Standards (NRC 1996) and your school's curricular objectives, and

■ lessons and assessments that bring about continuous instructional improvement and student achievement.

We should add that we don't recommend designing your entire curriculum over the summer. Good science is most often achieved through

Topic: Classroom Management
Go to: www.SciLinks.org
Code: DGS35

small steps and "subtle shifts," as San Francisco's Exploratorium calls them, guided by an overall plan. The key concept here is recognizing that much of the accomplishment attributed to master middle school science teachers stems from their preparations for the opening of school, which, in turn, are founded on their expectations for themselves and for their students.

In a learner-centered classroom, the teacher must be aware of the range of needs her students will bring to class on those first days. Middle schoolers start the school year in varying degrees of preparedness to learn, according to their home and family situations, their health and nutrition, their attitudes toward teachers and schooling, and their previous experiences with science as a subject. In the special case of students who are moving to middle school for the first time—where they will switch to a departmentalized class schedule with five or more

These are the seven things students want to know on the first day of school.

Source: Wong, H. K., and R. T. Wong. 1998. *The first days of school: How to be an effective teacher.* Mountain View, CA: Harry K. Wong Publications. © Harry K. Wong Publications. Reprinted with permission.

teachers, bell schedules, and passing time between classes—a number of basic uncertainties arise, as illustrated in the cartoon on this page.

The answers to the questions asked in the cartoon, among others with which students begin the year, shape student expectations, attitudes, and behaviors. Their questions must be carefully considered due to the safety concerns, student dynamics, and the interactive nature of good science in the middle grades.

Classroom Management
Management vs. Discipline

This is the section in which we connect the dots, so to speak, between the traits and needs of the middle school student described in Chapter 1, the characteristics of inquiry-based science instruction discussed in Chapter 2, and the procedural strategies teachers use to successfully channel bursts of adolescent energy in a meaningful direction. It can be done!

At the outset, we need to make a distinction between *discipline* and *classroom management*. In our experience, *discipline* almost always entails reacting to disruptions and then assigning consequences, usually negative; *classroom management* is about planning and preparing procedures to maximize student time-on-task and productivity. The most effective middle grades teachers structure their lessons—from the pivotal first few days of class—using procedures that students learn and internalize as routine.

This is not to say a teacher shouldn't have rules, which are what most teachers associate with "discipline"; indeed, a limited number of rules are an integral part of the classroom management plan. A middle school teacher who is a "disciplinarian" tends to post a list of rules—or sometimes may involve students in writing the rules and then posts them—as an accountability measure: "The rule is posted up there [teacher points with an index finger]; therefore, if you break it, you knew better and deserve what you get." In this sense, the rules are about deterrence—punishment—and are not integrated into the learning continuum.

An effective classroom manager may post the same rules—literally—and may or may not involve students in determining what they should be. The key difference is that a classroom manager teaches the rules to the students over the course of the first week or so of school, in conjunction with other procedures for taking roll, training students to begin each class by sitting down and working on a short assignment posted on the board, breaking down labs with five to ten minutes remaining in class, and being dismissed (where appropriate) only when the teacher says so. "Discipline"—or fear of punishment—now becomes "classroom management"—a learning process that manifests as a practiced behavior. This is the power of teaching procedures.

Perhaps the cardinal rule governing the behavior and attitudes of middle schoolers in good science classrooms would be something like this:

> *As long as you act like a scientist when we do science, you will be treated like a scientist and enjoy many exciting investigations and discoveries. If not, you get to watch the rest of us do science!* (Variation: *You get to be a retired scientist!*)

This expectation has served us (and many of the teachers we consulted in writing this book) as a very effective management strategy with middle-level students.

In Table 5.1, we list a sampling of ineffective methods and effective procedures that illustrate the advantages of a well-planned approach to classroom management.

Procedures, and rules in particular, satisfy an adolescent's need for structure, safety, and predictability, as discussed in Chapter 1. Part of the objective when teaching about the rules is to make students aware of the need to follow procedures to avoid unfavorable outcomes, such as accidents. We're not suggesting that classroom management plans not feature consequences for poor choices. Indeed, the most effective plans include rules with consequences that logically fit the student behavior in question. Middle school students need, and on a certain level, *want* to know the limits, but they are also looking for evidence that the world, and adults, are basically fair. That is why we advocate "teaching the rules" rather than simply posting them like a skull and crossbones! Serious infractions and disruptions, and certainly any behavior that endangers another individual in the classroom, demand major but also reasonable consequences, and middle school students should be taught why the consequences follow—and fit—misbehavior in the science classroom.

In the best-managed middle school classrooms we've seen, rules, consequences, and management procedures are *taught*—and tested (or quizzed)—during the first week of school and throughout the year.

Some effective middle school teachers prefer to determine and post the rules without consulting their students; others use a more democratic approach. Either approach can be compatible with classroom management plans tailored to adolescent learners and to the inquiry-based approach, although the more-participatory systems help to develop a sense of classroom community that is common in the effective middle school classrooms we've known. Kohn (1996) asserts that community building in a classroom

TABLE 5.1

INEFFECTIVE METHODS VS. EFFECTIVE PROCEDURES FOR MIDDLE SCHOOL CLASSROOM MANAGEMENT

INEFFECTIVE METHODS	EFFECTIVE PROCEDURES
Let students sit where they like.	Assign student seating on day one.
Post rules on the wall; refer to them when students act inappropriately.	Teach and test classroom expectations in first weeks; reinforce throughout year.
Leave parents out of the loop until a situation escalates, course grade drops, and/or end of term nears.	Involve parents in classroom management with early positive communication and follow-up.
Neglect to properly educate students about what to do in the event of an absence, an emergency, a substitute teacher, a missed assignment, and unexpected "free" time.	Establish class procedures well in advance of absences, emergencies, substitute teachers, makeup assignments, and unexpected "free" time.
Take roll while students chat or sit idly.	Start class with a brief student task.
Begin teaching without outlining the lesson, expectations, or objective(s).	Post daily schedule, including start- and end-of-class activities (bell work), what's due, activity/lab, and task(s).
Announce assignments verbally, with little notice; no student accountability for recording assignments or completing or submitting them.	Post assignments in advance to be recorded and checked in by students on calendars they maintain (which are checked periodically by the teacher).
Correct inappropriate work habits as they occur; limit cooperative/lab activities based on poor behavior.	Teach students how to work independently and in groups, how to get your attention, how to clean up after lab or activity.
Respond inconsistently to requests to move or leave based on circumstances.	Train students how to move about or leave the classroom during seatwork and labs.
Respond to inappropriate questions or comments.	Teach students how to offer and respond to questions and criticism.
Give free time at end of class—e.g., allow students to congregate at door.	Instruct students how to end a class and let them know that you, not the bell, are in charge of when they can leave.

begins when teachers invite students to participate in finding solutions and developing rules and procedures; as Kohn notes, "Students learn how to make good choices by making choices, not by following directions" (p. 78). Kohn's position is especially compelling when considering the developmental levels of middle schoolers, who "are more likely to be better behaved when there is no need for them to struggle to assert their autonomy" (p. 81).

Beyond developing classroom unity, successful middle school educators strive to develop broader agreement when it comes to classroom management by collaborating with other staff to adopt schoolwide behavior expectations—for students *and* teachers. Such expectations can help to unite the independent groupings or schools-within-schools (sometimes called "tribes," "houses," or "teams") that are increasingly common in U. S. middle schools.

A Strategy to Avoid

There is a "discipline" strategy teachers sometimes employ that we strongly discourage. In keeping with the goal of building community (and in the spirit of developing relationships with students), we believe middle school science teachers should avoid publicly singling out or humiliating individual students for misbehavior. A better way can be as simple as asking a student to step outside briefly to avoid a public confrontation or to meet for a discussion at lunch, homeroom, or recess, depending on the circumstances of the situation and the need for immediate intervention. (A private conversation about expectations eliminates many peer pressure complications.) Often a calm, respectful request to confer one-on-one is all it takes to encourage a student to refocus his or her attention, and the lesson can continue with minimal disruption. Other times, it's not quite that easy—but remember that the teacher's objective should always be to build relationships and community rather than break down a student's will or self-esteem. This should be a guiding principle of a middle school teacher's classroom management efforts. Presence of mind, objectivity, and several deep breaths will pay off in the long run.

Rewards and Praise

Negative consequences are commonplace middle school disciplinary approaches. However, many middle-level educators also incorporate positive incentives such as praise and rewards into their classroom management plans. Is positive reinforcement an effective management strategy? The answer is, "It depends." Some experts argue vehemently against the use of tangible rewards such as candy and stickers. We are conditioning our children to expect a reward for certain behaviors, the argument goes, and reinforcing their need for extrinsic motivation, diminishing their intrinsic drive, and programming the next generation with a "'what's in it for me' welfare and bribery system" (Wong and Wong 1998, p. 163). According to Kohn (1993) "The more rewards are used, the more they seem to be needed" (p. 17). Meanwhile, teachers have become so dependent on doling out rewards as incentives that they lack other, more creative motivational strategies.

Some forms of rewards are effective and appropriate in the middle grades. Our experience and some studies (e.g., Marzano, Pickering, and Pollock 2001; Jensen 1998) indicate that positive reinforcement in the form of verbal praise is actually more effective than tangible rewards such as candy. Praise applied appropriately and sparingly appears to have a positive impact on student attitudes and behaviors when it is contingent on students having reached a set standard or expectation.

Acknowledging Differences

Also in regard to classroom management, we urge teachers to be sensitive to the behavior patterns of children from cultural or demographic backgrounds that are different from their own. When teachers fail to take into account such differences, they can develop management procedures that are ineffective or that provoke a very different reaction than the teacher intends.

Are we advocating different rules for different students or that teachers apply consequences inconsistently? Perhaps this question is best addressed by asking another question: Is the goal of classroom management in the middle school to achieve "equal" treatment of all students? Our answer is no, because, as we discussed in Chapter 1, an important emotional need of middle-level learners is to be treated as individuals, with the teacher showing sensitivity toward the specific circumstances and emerging identity of each student. In fact, when it comes to disadvantaged students, who as adults are profoundly underrepresented among the ranks of science teachers, scientists, and scientific professionals because proportionally very few of them attend college, we are eager to discover and pursue new ways that will help all students succeed. In this regard, we direct readers to the important work of Ruby Payne (1998), whose research and training for educators on strategies for teaching children living in generational poverty is extremely useful for at-risk school communities in general. The future of good science—inclusive of diverse backgrounds and experiences—depends on this sort of resourcefulness.

Lesson Planning for Good Management

Basic to good science teaching is establishing an atmosphere of "investigation fervor." Early in the year, teachers need to let students know that they *are* scientists and that in the course of the term they will use and reuse their skills in the class. What skills are necessary for students to think and act like scientists? First, teachers need to make students understand that scientific research always starts with a question or a puzzling observation that raises a question. This is a platform for introducing (or reintroducing) students to a "scientific method" they can employ to conduct investigations. Sample focus questions that teachers can use to generate and guide inquiry activities are found in Chapter 7.

Second, teachers must introduce students to the process skills used in science:

- observing
- collecting data
- estimating
- problem solving
- predicting, hypothesizing
- investigating
- measuring
- classifying
- building models
- making graphs
- controlling variables
- discovering or determining cause and effect
- making inferences
- communicating
- drawing conclusions

Teachers should design lessons to activate these skills while helping students become more comfortable asking questions, defending or justifying their answers, and generally being skeptical of results and conclusions. These process skills are generic and applicable to any topic a teacher may choose to teach within the framework of the National Science Education Standards. Providing students with lab settings that maximize use of the greatest number of these process skills will expedite their competency and assist them in becoming scientists.

Teaching Safety

Middle schoolers are by nature incredibly curious young people. This trait, combined with their tendency to explore the limits of appropriate behavior, forms a potent recipe for science disasters. Consequently, no discussion of classroom management in the context of middle-level science education is complete without considering the issue of student and teacher safety. Safety is primary to good science—everything else is secondary.

Hands-on science experiences present more risks than those associated with reading textbooks and completing worksheets. The professional judgment of the teacher is the most important factor in determining which activities should be used and which activities should be omitted. Accordingly, science safety is maximized by teachers who set clear expectations for student behavior.

As with all management expectations, safety is a matter of teaching, testing, and reinforcing procedures. Questions about safety, particularly issues surrounding lab activities, should appear in the start-of-term student preassessment in order to determine the extent of the students' previous safety preparation.

In addition, prior to engaging in any science activities, students must be required to complete and return a safety contract signed by their parents and also score 100% on a safety quiz. (Readers may refer to the Flinn Scientific Web site at *www.flinnsci.com* for a sample contract and quiz.) We urge teachers to include a specific clause in the contract that warns against unauthorized experimentation—good science is fun, but teachers are professionals and students should *not* try this at home!

Key procedures that we recommend teaching and implementing from the start of the term are the following:

- Make it clear that safety is the highest priority in your classroom and that students who choose not to behave like scientists during an activity or project will observe, rather than participate, for that day.

- Familiarize and train students regarding location and use of sinks, eyewash, chemical shower, first aid kit, and other emergency stations.

- Assign a materials manager for each group who inventories items before and after the activity.

- Designate a signal to indicate cleanup and inventory time.

- Establish an expectation for cleanup and materials inventories so that students understand class won't be dismissed until the room is in order and all supplies are accounted for.

- Make sure that materials are organized and packaged in a box or tub for each group.

- Locate materials so students don't have to carry them too far (how far is "too" far depends on the hazards associated with what is carried).

Topic: Safety in the Laboratory
Go to: *www.SciLinks.org*
Code: DGS41

■ Arrange the physical space in the classroom to accommodate the traffic patterns of the specific lab activity, including access to safety equipment, as well as setup and cleanup considerations.

■ "Declutter" prior to lab days, as clutter compounds the possibility of accidents and injuries.

> **THERE IS
> NO SUBSTITUTE FOR SAFETY!**
> 1. Train your students in safety and emergency procedures.
> 2. Provide equipment in good operating condition.
> 3. Supervise students AT ALL TIMES!

A lab safety checklist, which can be posted in the room following the teaching of a safety unit at the start of the year, might look like Figure 5.1. (See also the lab safety rules in Appendix D.) Teachers should review every item, with every student, before every lab.

We celebrate inquiry and are excited when our students are interested enough to pursue their investigations. However, over the years and increasingly, very dangerous "science" activities are available to young people with access to the Internet and other sources. Years ago, this prompted the "See Me" rule, which is as follows:

> *If you want to try your own experiment, please SEE ME first! We can investigate it together, and I will evaluate the safety of the proposed experiment—bring instructions or information if you have any. If it is safe, you may proceed with the investigation under my direct supervision.*

After the first few times we explained the rule, we'd ask in a lab write-up or quiz for students to repeat it. This helped emphasize the

FIGURE 5.1
LAB SAFETY CHECKLIST

BEFORE YOU BEGIN

1. Do you know who your lab team members are?

2. Do you know the task for which your lab team is responsible today?

3. Do you know the procedures for teamwork?

4. Do you know proper safety precautions needed for today's lab (goggles, apron, gloves, etc.)?

5. Do you know whom to contact in case of a classroom emergency?

6. Do you know where basic safety supplies (baking soda, eyewash, water) are kept in case you are asked to locate them?

7. What is the *first* thing to do in the event of an emergency?

AFTER THE LAB IS COMPLETED

1. Did your team work well together?

2. Did your team complete the assigned task?

3. Did your team follow all the safety rules?

4. Did your team need to use any first aid supplies?

5. If yes, which ones and why?

importance of open communication between teacher and students and reiterated the gravity of the expectation.

We also made sure the parents understood "See Me." For added safety, we recommend that teachers require all independent investigations to include a complete description of the proposed inquiry with verifiable signatures of the parent(s) indicating they understand that all procedures must take place at school under the teacher's supervision. (There have been cases where teachers were held liable for student "science experiments" at home when the teacher talked about specific subject matter in class and neglected this sort of explicit caution.) Further, we advise that if a teacher is inexperienced or unfamiliar with the specific area of a student's independent study interest, it's essential to seek the advice of experts to help evaluate the safety and educational value of the student's proposal.

An Emotionally Safe Environment

Beyond a conscious concern for safety in the classroom, the environment should be tailored to the needs and interests of students and teachers who have to live together in the room for months. We've seen a wide variety of middle school classroom environments that suited student needs and were associated with successful students and teachers year after year. Generally speaking they were places with some color, warmth, and public acknowledgment that students are important to the teacher (e.g., student work posted, birthdays listed). Some teachers find that creative and varied desk arrangements facilitate different instructional activities better (and can cause a fairly marked difference in how students behave and interact). In Chapter 1 we saw that middle schoolers need a blend of spontaneity and structure in a teacher's instructional methods; this is probably a good rule of thumb concerning the class environment, too.

In sum, we agree with Rick Wormeli's call for creating "an emotionally safe environment" (2001, p. 8), focusing on a teacher's role of empowering learners in the middle grades:

> There are many ways to boost the confidence levels in our middle school classrooms without getting lost in self-esteem hoopla such as putting up "happy" posters. Be pleasant to students. Call them by their first names. Greet them at the door. Smile often. Catch them doing something well. Crack a few jokes. Ask questions that show your interest. Applaud risk taking. Share excellent homework or test responses with the rest of the class. Allow occasional democratic voting in the class. Refer one child who is an expert on something to another child who needs help, and make sure you rotate the expert's role. Ask students to tutor their peers after school. Give them responsible jobs in the classroom. Ask them to serve as hosts for guest lecturers. Point out moments of caring among peers that occur in class. Post their accomplishments in class. Make at least one positive phone call or note home for each child per year. (pp. 8–9)

Efforts such as these will go farther, based on our experience, than any number of pop culture icon posters or slogan banners to win the loyalty and interest of middle school learners.

References

Jensen, E. 1998. *Teaching with the brain in mind.* Alexandria, VA: Association for Supervision and Curriculum Development.

Kohn, A. 1993. *Punished by rewards.* Boston, MA: Houghton Mifflin.

Kohn, A. 1996. *Beyond discipline: From compliance to community.* Alexandria, VA: Association for Supervision and Curriculum Development.

Marzano, R. J., D. J. Pickering, and J. E. Pollock. 2001. *Classroom instruction that works: Research-based strategies for increasing student achievement.* Alexandria, VA: Association for Supervision and Curriculum Development.

National Research Council (NRC). 1996. *National science education standards.* Washington, DC: National Academy Press.

Payne, R. K. 1998. *A framework for understanding poverty.* Highlands, TX: RFT Publishing.

Wong, H. K., and R. T. Wong. 1998. *The first days of school: How to be an effective teacher.* Mountain View, CA: Harry K. Wong Publications.

Wormeli, R. 2001. *Meet me in the middle: Becoming an accomplished middle-level teacher.* Portland, ME: Stenhouse.

Ten Activities for Middle School Science

Developmentally Appropriate, Inquiry- and Standards-Based

Before we move to the activities, we need to give a disclaimer or two concerning the danger of using any "template" for lesson planning. The components of the lesson-plan template presented in these activities are all integral to inquiry-based science instruction, but, as teachers know, units and lessons need to be tailored to the specific needs of different groups of students, different days, and even variations in unit objectives. Teachers using our template are encouraged to modify it as needed to fit their own circumstances. (We offer suggestions for making modifications in Chapter 7.) What's more, we know that the activities might look cumbersome, but we've opted to provide extensive detail so that they are truly ready-to-use.

Let us reiterate that inquiry-based lessons should be designed with practicality and instruc-tional effectiveness in mind. It is frankly not feasible to do inquiry activities all the time, in part because the setup, process, and cleanup are time intensive. Neither is it prudent from a pedagogical standpoint. Science instruction should be varied—structured to incorporate a range of strategies that will engage different learning styles.

Also, readers should be aware that although the activities in this chapter are offered as ex-amples of inquiry-based instruction, they are in no way intended to represent a complete middle school science curriculum for any particular state or school district. Our activities emphasize the National Science Education Standards (NSES) (NRC 1996) content areas of (a) inquiry, (b) unifying concepts and processes, and (c) the nature of science—content standards that make up the foundation of conceptual understanding in middle school and provide a context for further study of, for example, biology, geology, chemistry, physics, and astronomy. Different states have varying expectations for what K–12 teachers will cover in the traditional content fields of life, Earth, space, and physical sciences; we selected inquiry- and process-focused activities that adhere to the NSES content areas and can be used in multiple state-standards contexts.

Content and Kit-Based Instruction

While most states have specific science content standards, generally emphasizing the traditional content areas, the NSES content standards provide a framework for teaching core science concepts. The eight categories of content standards are

- *Unifying concepts and processes in science*
- *Science as inquiry*
- *Physical science*
- *Life science*
- *Earth and space science*
- *Science and technology*
- *Science in personal and social perspectives*
- *History and nature of science* (NRC 1996, p. 104)

Increasingly, state science standards and district curricula reflect these NSES content standards and a corresponding emphasis on inquiry, and school systems around the country are using science kits in their inquiry-based programming because they remove a number of roadblocks to implementing inquiry.

We define *kits* broadly, to include a range of products from district- and school-developed materials to commercially produced curricula. Kits contain materials necessary for students to conduct scientific investigations and experiments organized into units spanning several weeks. The units usually consist of 10 or more lessons and are structured according to teacher guides. While many educators and some school systems develop and refurbish their own kits, units for middle school instruction may also be purchased from a number of vendors.

Kits began as a not-for-profit venture on the part of teachers who wanted to share an excellent lesson and boxed the self-contained student activities. Subsequently the demand for kits led to their creation and distribution by curriculum marketers.

The commercial kits claim to be aligned with the NSES, and our experience supports that claim. Many units incorporate the use of science lab notebooks and frequently integrate math, reading, and writing activities. At local workshops and large conferences, kit vendors train teachers to become familiar with unit activities and to lead students through the materials with a minimum of direct instruction. (See Chapter 8 for information about vendors of kit-based instruction).

Generally speaking, kit-based science instruction is readily compatible with the instructional framework established in the NSES, and kits can facilitate hands-on instruction. However, it's important for us to point out that good science isn't *in* the kit; it emerges *from* the kit. That is, beyond using the engaging activities contained in a boxed science unit, the teacher is responsible for framing the lesson with focus questions, guiding the inquiry as a coinvestigator, helping students make connections to important concepts and key terminology, and constantly monitoring learning to assess the need for reinforcement, review, and enrichment.

Activity Template

There are many possible variations of our 10 sample activities, and we certainly don't attempt to illustrate all of them in this chapter. However, we do use a template that can be used with a variety of inquiry-based lessons. Our template is structured to be adapted to a "typical" guided-inquiry lesson (see p. 1 for a discussion of *guided inquiry*)—challenging students, according to the 5E model, to engage, explore, explain, expand, and evaluate (Bybee 1997; see pp. 18–19 for a discussion of the model). Below we describe each section of our template.

Standards

This section identifies which national science (NRC 1996), math (NCTM 2000), and literacy (NCTE/IRA 1996) standards are explicitly met by the activity. (This would be the section where teachers could also identify which state standards are accomplished by the activity.)

Integration

This section identifies the subject areas represented by the tasks in the activity. It demonstrates the connection between science, math, and literacy found in most good science activities.

Objectives

This section defines what a student will be able to do as a consequence of successfully accomplishing this activity.

Key Words

This section lists and defines the words that students should be familiar with before undertaking the activity and the words that they will learn during the activity.

Focus Questions

Focus questions can be used as a form of anticipatory set, as a way to generate discussion with the whole class, or as "tasks" assigned to small groups or teams to initiate investigations. The inquiry process starts with a question or problem. Sometimes it's a question or problem posed by the teachers; other times it's a question or problem that students generate or encounter themselves.

Background

This section provides information teachers can use to help them when preparing for the activity, generally going beyond what students need to know.

Preparation and Management

This section provides the nuts and bolts of the activity. In addition to the prep time and the teaching time, this section gives the following:

- *Materials:* This section lists the items that are needed to complete the activity.

- *Procedures:* This section provides the guiding questions, demonstrations, or direct instruction the teacher needs for the activity. It details the steps for conducting the activity. (Guiding questions are an extension of "guided inquiry" and are used to assist students as they carry out their investigations. With middle schoolers, teachers need to be prepared to use questions that help students bridge concepts or that nudge students toward clues and clarifications that will help their investigations to be more productive, without giving away the "answer.")

Discussion

Questions are especially important in inquiry-based activities. Students should be engaged in challenging and questioning their hypotheses, processes, and findings and those of other students. The discussion should not be filtered through teacher responses and commentary.

Extensions (application and inquiry opportunities)

This section directly reflects the 5E (Bybee 1997) concept that underlies the activities. Teachers should devise ways (and challenge students to figure out how) the concepts in the activity can be demonstrated in real-life contexts—for example, at home in the kitchen, on the playground, or in the school gymnasium. Extensions can include investigations in which students pursue questions of their own that arise from the unit or lesson just completed (with teacher awareness and permission).

Assessment

The goal of assessing an inquiry activity is to determine (1) whether students learned the concept(s) in the lesson and (2) how instruction might be altered to better promote student un-

derstanding. The assessment may take one of many forms, including pictorial representations, student response, performance of a task, teacher observation, or rubric, depending on the learning to be measured—and on the creative talents of the teacher!

Activities

The following activities have been successfully used for years by teachers all over the country, working with a wide range of students of different ages, demographics, and ability levels. We've modified and adapted them to suit the needs of our students, just as you will. (There are many resources listed in Chapter 8 that will lead you to more lesson ideas.)

Before moving to the activities, we would like to reiterate how important it is that the teacher provides adequate safety instruction and procedures for students to follow and closely monitors students who are conducting the investigation. (See Chapter 5, pp. 41–43, and Appendix D for more on safety issues.)

Thinking Like a Scientist

All kids are scientists—so let's invite them to show it. This activity shows how students can make sense of lab reports written by someone else. Like real scientists, students get to analyze data, draw conclusions, and make inferences from what is before them. After these skills are in place, students begin the process of replicating what others have done and then support or refute the results of other scientists' work. This activity lets middle schoolers experience good science by doing it.

Standards

Science (NRC 1996)	Inquiry, Nature of Science
Math (NCTM 2000)	Data Analysis and Probability
Language (NCTE/IRA 1996)	Comprehend, Interpret, Evaluate, and Appreciate Texts; Research, Pose Problems, Gather and Evaluate Data, and Communicate Findings; Use Language for Exchange of Information

Integration

Science:	Data analysis, scientific thinking
Math:	Data analysis
Language:	Reading, viewing, and presenting

Objectives

- Students will become familiar with seven science process skills—problem solving, discovering/ determining cause and effect, making inferences, drawing conclusions, classifying, predicting/ hypothesizing, and building models.

- Students will be able to read, decode, and analyze the written graphs of other scientists.

- Students will be able to evaluate lab reports for reasonableness and logic by asking the following questions: Does the report make sense? Does it follow the logical pattern of the data from the experiment? Could I explain it to someone?

Key Words

logic—the use of strict rules of reasoning to show whether a statement is true or false

metacognition—awareness or analysis of one's own learning and thinking processes

replicate—to duplicate or repeat

Focus Questions

■ "What does the data say?"

■ "What can be proven and what can't? Does data 'prove' anything?" (Here the teacher is working toward the concept that scientific data can be used to refute or support, but not to "prove.")

■ "What is fact and what is fiction?"

Background

Basic to good science teaching is establishing an atmosphere of investigative fervor. Early in the year, let students know that they *are* scientists and that you plan to use and reuse their skills. Skills that will pay the highest dividends and that will be celebrated are those that allow students to think like scientists and act like scientists. Those science process skills are as follows:

■ Problem solving

■ Discovering or determining cause and effect

■ Making inferences

■ Drawing conclusions

■ Classifying

■ Predicting/hypothesizing

■ Building models

Students should become familiar with the fact that research always starts with a question or a puzzling observation that raises a question. This understanding will become more sophisticated as students mature. Additionally, students should become comfortable with asking questions, defending or justifying their answers, and being skeptical. These skills are generic and applicable to any topic you teach. By providing your students with lab settings that maximize the use of the greatest number of process skills, you are helping them to become true scientists. Teach 'em to think!

Preparation and Management

■ *Prep time:* 15 min. Make copies of previous lab reports written by students you have had in the past. If you're a new teacher, borrow lab reports from colleagues. Be sure to remove all student names from the reports. These reports should be of high quality. (In the Extensions section, students are given reports with vague or inconclusive analyses to evaluate.) Be sure to save what your students create in labs this year to use with future classes in this activity.

- *Teaching time:* Two 45 min. periods. You may want to use one day to have students analyze data and a second day for their presentations.

- *Materials:* One copy of a completed student lab report from a previous year, per group of two to three students. You may want each group to have a different lab report to analyze.

- *Procedures:* Distribute a copy of a completed lab report to each group. (See Figures 6.1 and 6.2 for a completed lab report and a rubric that can be used to evaluate it.)

Ask the group members to review the work as scientists would. They should ask themselves:

1. Am I able to interpret the information on this document?

2. Does it make sense?

3. Can I replicate what the previous science team has done?

4. Do I agree with their conclusion(s)? Why or why not?

(continued on p. 53)

FIGURE 6.1

SAMPLE LAB REPORT: BUILDING AND FLYING A PAPER AIRPLANE

QUESTION
How can we make a paper airplane travel farther?

HYPOTHESIS
If we add weight to our plane, then it will fly farther.

PLAN
1. Fold paper into shape of a plane.
2. Fly it without a paper clip.
3. Add one small paper clip to the bottom center of the plane.
4. Fly plane multiple times marking distances.

TEST
Without paper clip

Trial 1	Trial 2	Trial 3	Trial 4	Trial 5	Average
82 ft	50 ft	93 ft	32 ft hit ceiling	80 ft	67.4 ft

With a paper clip

Trial 1	Trial 2	Trial 3	Trial 4	Trial 5	Average
60 ft	74 ft	88 ft	62 ft	90 ft	74.8 ft

CONCLUSION
Adding the paper clip does increase the distance of the flight. We need more space to fly!

FIGURE 6.2
RUBRIC FOR EVALUATING
"BUILDING AND FLYING A PAPER AIRPLANE"

1. Did the team doing the test follow a procedure?

| 1 | 5 | 10 |
| ☹ | 😐 | ☺ |

2. Did the test seem appropriate for the hypothesis?

| 1 | 5 | 10 |
| ☹ | 😐 | ☺ |

3. Did the test seem fair?

| 1 | 5 | 10 |
| ☹ | 😐 | ☺ |

4. Did the students conduct enough trials? Did they gather enough data?

| 1 | 5 | 10 |
| ☹ | 😐 | ☺ |

5. Did the conclusion seem logical?

| 1 | 5 | 10 |
| ☹ | 😐 | ☺ |

6. Overall, how would you rate this team on their ability to think like scientists?

| 1 | 5 | 10 |
| ☹ | 😐 | ☺ |

(continued from p. 51)

Tell the students: "Be prepared to share orally with the whole class. You will be asked to tell the class what your report is about and what reactions different members of your group had to the report. All members of the group must participate in the presentation."

Discussion

Rich discussion can emerge from the presentation of information. Students should be encouraged to ask questions, challenge results, or offer alternative interpretations.

Extensions (application and inquiry opportunities)

■ Give students the opportunity to apply their scientific thinking by developing several investigation options, preferably extensions of activities completed in class, and allow them to choose one and carry it out.

■ Give the groups a new lab report with a vague or inconclusive analysis of the lab. Have each group discuss the interpretation, then make a list of facts that will support or refute the interpretation. Set up a debate between groups to argue in support of or in refutation of the interpretation of the data, based on the information in the lab report.

Assessment

■ As above, teachers can use copies of lab reports from previous years (with student names removed). First-year teachers can borrow some reports from a colleague, or "invent" a few. Make sure that each member of the group has his or her own copy of the report. When students complete a lab report evaluation, have each group trade it with another group. (Each report is passed only once.) If the report is done clearly and correctly, the new group should be able to follow the report and concur with it or refute it. If it is not written clearly, members of the receiving team may ask questions in writing or orally to clarify any uncertainties. Sometimes several groups run into the same questions; in that case, the teacher can address the questions with the whole class. This is a great way to assess your teaching as well as student learning.

■ Another assessment option is to have each group present and defend their analysis of the report.

2 *activity*

Attributes

Can you believe your eyes? Can you describe items in great detail? How good are you at seeing (and hearing and smelling) what's really there? This introductory-level lab encourages students to think like scientists by distinguishing factual characteristics from their opinions or preconceptions. Students enjoy this activity and are surprised to discover that their senses and preconceived notions can deceive them—as in our "Extensions" variation, when we present them with cookies loaded with aromatic spices instead of sugar!

Standards

Science (NRC 1996)	Inquiry, Nature of Science
Math (NCTM 2000)	Data Analysis and Probability; Measurement, Reasoning, and Proof
Language (NCTE/IRA 1996)	Research, Pose Problems, Gather and Evaluate Data, and Communicate Findings; Use Language for Exchange of Information

Integration

Science:	Science observations
Math:	Development of mathematical terminology, measurement
Language:	Written records

Objectives

■ Students will practice observations and analysis skills.

■ Students will distinguish between fact and opinion and will write descriptions of what they observe.

Topic: Senses
Go to: *www.SciLinks.org*
Code: DGS54

Key Words

attribute—a quality, characteristic, or property of a person, thing, or group
observation—the act of watching something closely and recording how it behaves or changes under certain conditions

Focus Question

"How do you describe an object without naming it?"

54

National Science Teachers Association

Background

Before doing science experiments in your class, it is helpful to build in time for developing skills that scientists use. You want your students to think like scientists, act like scientists, and eventually *be* scientists! Observation of physical characteristics (properties) of objects is essential to doing science.

Preparation and Management

- ■ *Prep time:* Minimal—maybe 2 min.

- ■ *Teaching time:* 45 min.

- ■ *Materials:* Any common classroom object, such as a pencil, ballpoint pen, or board eraser

- ■ *Procedures:* Hold up the common classroom object. Ask students to study—to *observe*— the object you are holding. Then ask students to name attributes of the object. Students should be careful not to name the object and should be as factual as possible. A student might say "small," "pointy," "lightweight," or "soft." Challenge students to explain and defend their descriptions—"Is that something you can prove, demonstrate, or measure based on what you can now observe?"

Discussion

Ask other class members if they agree with the attributes a student has named. Why or why not? Encourage discourse among students as opposed to allowing students to direct all comments to you. Insist on professional courtesy and respect in the exchanges among and between scientists!

Extensions (application and inquiry opportunities)

This activity can be varied by using homemade cookies, with some having ingredients (large amount of pepper, aromatic spices, food coloring) added to change the preconceived notion of what the students are observing and to provide different results for different groups. (**Safety Note:** The cookies should be observed but not tasted. Remind students that no eating or tasting should occur in a science classroom or lab; also, caution students to never taste an unknown substance.)

Assessment

- ■ Distribute a variety of mystery items in paper lunch bags. Select objects with a variety of shapes, weights, densities, and textures. Ask students to write a complete description of the items listing as many attributes as possible. If students are able to write a complete description, other students may be able to identify the item without even seeing it.

- ■ 20 Questions: The teacher holds up a limited number of common objects. Students choose whichever they wish and write a description. Several students are chosen to be, in turn, the object of "20 questions." Class members are allowed to ask specific "closed" questions relating to attributes ("Is it larger than a pencil, or smaller?") until they guess the object. (A closed question can be answered with a single word [such as yes or no] or a short phrase.)

Penny Water

"Penny Water" is a fast, fun, material-simple way to raise the issue of variables and the need to control them. In "Penny Water," it's easy for students to see and identify variables, and the activity reinforces the importance of conducting a "fair test." "Penny Water" is a great introduction to scientific processes and guided inquiry because students are free to conduct an investigation based on a teacher-generated question. The *NSTA Position Statement on the Nature of Science* (Appendix C) supports the argument that there is no "single" scientific method—you be the judge!

Standards

Science (NRC 1996)	Inquiry, Nature of Science, Unifying Concepts
Math (NCTM 2000)	Data Analysis and Probability; Algebra/Patterns, Reasoning, and Proof
Language (NCTE/IRA 1996)	Research, Pose Problems, Gather and Evaluate Data, and Communicate Findings; Use Language for Exchange of Information

Integration

Science:	Scientific processes, experimentation, variable
Math:	Mean, median, mode, range
Language:	Record and organize data

Objective

Students will be able to use scientific process skills—problem solving, discovering or determining cause and effect, making inferences, drawing conclusions, classifying, predicting, and building models—to solve a problem.

Key Words

mean—a number that is midway in value between other numbers; the average

median—(1) the middle number in a sequence of numbers listed from smallest to largest if there is an odd number of numbers. In the sequence 3, 4, 14, 35, 280, the median is 14. (2) the average of the two middle numbers of a sequence of numbers listed from smallest to largest if there is an even number of numbers. In the sequence 4, 8, 10, 56, the median is 9 (the average of 8 and 10).

mode—the value that occurs the most in a set of data. In the set {25, 40, 72, 64, 40, 10}, the mode is 40.

predict—to foretell on the basis of observation, experiences, or scientific reason

range—the difference between the smallest and largest number in a set of data. If the lowest test score of a group of students is 54 and the highest is 94, the range is 40.

reliability—the extent to which an experiment, test, or measuring procedure yields the same results on repeated trials

validity—the extent to which a study or test measures what the researcher says it measures

variable—a part of a scientific experiment that is allowed to change in order to test a hypothesis

Focus Question

"How many drops of water will fit on the head of a penny?"

Background

Students will see that the test featured in this activity is not "fair" and that some comparisons are not valid. This activity is an excellent way to show students the presence of variables and the scientist's need to control or manipulate them. Students have no trouble identifying the following variables: different pennies, the age of the penny, the wear of the penny, the surface on which the experiment is conducted, the difference in eyedroppers, the technique used to squeeze out a drop, the temperature of the water, the temperature of the penny, and which side of the penny is used (heads or tails). Students will even notice and mention the city in which the pennies are minted!

Note about content: This activity is about variables. The science behind why a coin will hold so much water without spilling over is an entirely separate lesson addressing cohesion and adhesion and surface tension. If students bring up these scientific processes, by all means address the questions, but don't make it the focus of this lesson. "Penny Water" is intended to illustrate variables and how to work with them.

That said, for those of you who just *have* to know something about the concepts at work in the activity beyond an understanding of variables before you will feel comfortable doing this activity, here is a general explanation. Please understand this is for you as the teacher so that you will have enough background to minimize the potential for misconceptions by your students. So here it goes:

The water molecule has an angular (bent) shape. This is the result of two nonbonding pairs of electrons on the oxygen atom. According to the valence shell electron pair repulsion (VSEPR) theory, these nonbonding pairs of electrons are the reason there is a bent shape, and a large relative space of negative charge. The other single electrons are shared with the single electron on each of the hydrogen atoms. The single electron on each hydrogen atom is pulled closer to the oxygen atom than to the hydrogen atom. Since the hydrogen atom is composed of one electron and a nucleus of one proton, the proton nucleus of the hydrogen atoms causes a more positive space around the hydrogen atoms in the water molecule.

Therefore, the water molecule has a bent shape with a negative region around the vertice of the angle, and a positive region around each of the hydrogen atoms. When many of these molecules are close enough to be in a liquid state (liquid water), the molecules tend to stick together like magnets. The hydrogen ends (positive) are attracted to the oxygen vertice (negative end). This electromagnetic attraction causes liquid water to demonstrate the observed phenomena called *surface tension*. This is also why water can be siphoned, and capillary action exists. Anyway....we suggest you stick with variables as the focus of this activity.

This activity can be done formally with a lab report by each student or as a classroom exploration with each student taking notes about his or her learning and then preparing a KWL chart based on the experience. (A KWL—**K**now, **W**ant to Know, **L**earned—chart is typically displayed on a classroom wall and used periodically as part of the introduction to a unit or activity. It lists three basic questions to be addressed by students: (1) What do we already know or think we know? (2) What do we want to know? (3) What did we learn or discover?)

Some key ideas: Scientific processes are repeatable and are supported by the collection of empirical observations and data. Consistency of results supports the strength of the procedures and conclusions.

Preparation and Management

■ *Prep time*: 15 min.

■ *Teaching time:* 45 min.

■ *Materials (per group of 2 students):*

4 eyedroppers (a variety of sizes)

1 C water

3 paper towels

4 pennies

■ *Procedures:*

1. Pose the question "How many drops of water will fit on the head of a penny?" and have students predict what the answer will be.

2. Then ask them how they can find out "the" answer. Ask them how many times they will need to conduct the test (count water drops they place on a penny) to obtain an accurate answer. Encourage students to repeat the test at least five times.

3. After students conduct the tests in groups of two and record their results in their lab notebooks, have them analyze their data and try to make sense of it. Can they determine one exact answer? How?

4. Conduct a class discussion. Have some groups share their results. Record the raw numbers on the board or on an overhead.

5. Put the numbers of drops in order from smallest to largest. Lead students to and through a discussion of mean, median, mode, and range. Allow each group of students time to

arrange their results in order and to determine mean, median, mode, and range with their data. Ask them to compare their results to their predictions. Were they close? How far off were they? If asked to predict again, what would they say? Would they be closer? Why or why not?

6. Discuss the difference between a "wild" and an "educated" guess. How can we develop a pattern of thinking or a process approach that is generally accepted? Posting and referring to the seven steps of scientific processes (cited in the "Objective" section of this activity) at this point will be much more relevant to students than if they see the steps "cold." Referring back to the processes after students have seen the phenomena in action will add relevance to the activity and enhance consistency and reliability in later lab activities.

Discussion

Discussions should be conducted throughout the activity. Ideas to elicit from students and to reinforce include the concept of a "fair" test, identifying variables (which ones? Let the kids tell you, and refer to the "Background" section above).

Extensions (application and inquiry opportunities)

Have each group report their median number to a class data sheet (Figure 6.3). Analyze these results and arrive at a conclusion (whole-group discussion).

FIGURE 6.3
SAMPLE CLASS DATA SHEET FOR "PENNY WATER"

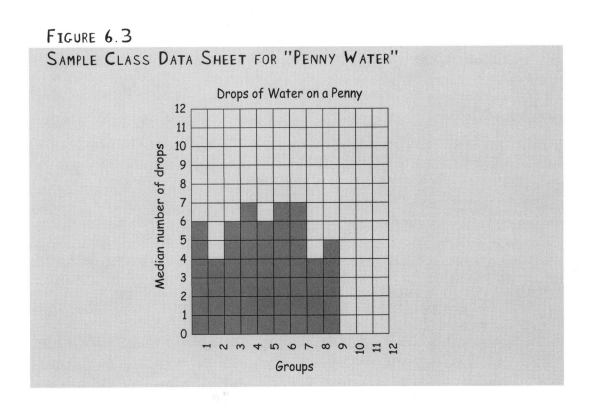

Assessment

■ Repeat this activity using other coins (nickels, dimes). Ask students to list variables involved and how to control them to influence a fair test (e.g., same size droppers). You could give them a description of a test involving different variables—some of which would be viable and measurable, and others not—and then let them evaluate the test.

■ Have the students set up the "rule book" for and conduct a fair test to see which group can get the most drops of water on a penny. (The students will control variables if they've learned from the activity—same penny, same water, same dropper, and so forth.) The winning group will be able to get the most drops on the penny *and* explain why their technique worked.

The Incredible, Edible Candle

What's in a candle? We all make assumptions . . . . This activity shows students how relying on prior knowledge can lead them to make errors in scientific observation. It is a real eye-opener to students when they are forced to reserve judgment until they have collected more data. This in turn encourages students to be skeptical about results until they are justified.

Standards

Science (NRC 1996)	Inquiry, Nature of Science
Math (NCTM 2000)	Data Analysis and Probability
Language (NCTE/IRA 1996)	Research, Pose Problems, Gather and Evaluate Data, and Communicate Findings

Integration

Science:	Science observations
Math:	Data analysis, vocabulary
Language:	Written records

Objectives

■ The students will recognize, describe, or measure physical attributes of objects (e.g., length, width, height, circumference, smell, composition, color, shape, weight, texture).

■ Students will recognize the difference between factual characteristics (big, tall, light) and opinion (fun, "cool," attractive).

Key Words

attribute—a quality, characteristic, or property of a person, thing, or group
skepticism—a doubtful attitude

Focus Question

"How could you identify and describe the object I am holding up?"

Background

Discrepant events are observations that do not match our previous knowledge and cause us to be uncomfortable and confused. This is an important aspect of science education. Students need to be skeptical of what they see and hear. Scientific thinking teaches them to question what they see, ask for proof, and reserve judgment until they have gathered data.

The success of this activity will hinge on whether the teacher has rehearsed. The best tip is to practice, PRACTICE, **PRACTICE!**

Preparation and Management

- *Prep time:* 15 min. to prepare potato candle. Prepared candle will begin to discolor after about 40 min. Using a thermos of cold water will keep the candle fresh for several hours prior to the demo.

- *Teaching time:* 45 min.

- *Materials:*

  1 large Idaho potato

  several slivered almonds (found in baking section of grocery store)

  apple corer or cork borer

  matches

- *Procedures:*

  1. Before class: Use the apple corer to cut a long cylinder of potato out of a large Idaho potato. Cut off both ends of the cylinder so you only see the cream-colored potato. For the wick, use an almond sliver. Shave off the outer cover of the almond and trim so it appears to be a candlewick. Make a small indentation in the top of the potato and stick in the "wick."

  2. So that students have to rely on visual cues, it's important to keep them observing at a distance as you display the candle. Ask students to identify the object (either as a class or individually—teacher's choice—they always fall into the trap! They say it's a candle.)

  3. Ask students to describe what the object looks like (color, shape, length, width, diameter, circumference, texture, composition).

  4. Turn off the lights and light the candle. Ask students to describe the candle further. What do they see now? (They might say, e.g., "light," "flicker," "colored flame," "translucence of the candle," "melting.")

  5. Have a student ready to turn on the lights as you blow out the candle. Ask students what they see now. (They will probably say, e.g., "The wick has changed color." "Smoke has been produced." "The candle smells." "The candle's shorter.")

  6. Compliment the students on their great observational skills. Review with them that they have identified the object and described its attributes (e.g., color, shape, odor, length) and

that they are very good scientists with great detective skills. Then casually take a bite of the candle, including the wick. (The wick will have cooled during the discussion.)

7. Students will be shocked and suffer a moment of disequilibrium. They will immediately want to know what the object is. Remind them that they told you it was a candle!

8. Students will begin to guess. On a T-chart (Figure 6.4) record their guesses. On the left side of the chart, record the guesses for what the candle is really made of. (Students will typically guess apple, potato, jicama, cucumber, pear, or turnip.) Record all guesses but do not confirm whether they are right or wrong. Ask students if each of those materials would burn.

FIGURE 6.4

T-CHART FOR "THE INCREDIBLE, EDIBLE CANDLE"

Students accept this activity, just taking it for granted it is a candle. Once they see the teacher take a bite of it, they are startled. They immediately accept that it is not a candle, but what is it? Below is an example of a T-chart to use with some typical guesses that students offer up for an explanation of what the teacher is eating

| WHAT IS THE TEACHER EATING? | |
CANDLE	WICK
Apple	Toothpick
Jicama	Paper
Turnip	Apple
Potato	Nut
Radish	

9. On the right side of the chart record the guesses for what the wick is really made of. They will frequently guess a match or a piece of paper. Remind them that you ate it. Would you eat a piece of paper? A match? Again record their guesses but don't confirm correctness.

10. If you will be conducting this activity more than once in the day, it will have maximum impact if you close each class with a solemn oath of scientific secrecy. We have found it helpful to tell students that they can tell others about the "really neat candle demonstration in science class." Beyond that, we admonish students to keep things quiet! We've found that our fifth graders have the most difficulty with loose lips, but it gets a lot easier to preserve the element of surprise with sixth graders on up.

11. NEVER TELL THE ANSWER! Remember, when you give the answer, all thinking stops. You want to keep your students thinking and problem solving as long as you can. This drives kids crazy at the start of the year, but they get used to it.

Discussion

Encourage students to challenge a theory shared by any student. Promote discourse between students through respectful questioning and debate.

Extensions (application and inquiry opportunities):

Have students research and demonstrate other discrepant events (see p. 62, Background), after clearing them with you for safety.

Assessment

Distribute one paper lunch bag to each group, with a common classroom or household object in it. (It is up to the teacher whether to use the same object in every bag or different objects in each bag.) Ask students to do the following: Look into the bag and make observations. Close the bag and try to write a description of the object in great detail without identifying what it is. Staple or tape the description to the outside of the bag. Pass it to another person in the class. The receiving student should be able to identify the object from the description on the outside (without opening the bag) if the attributes have been well defined and if the object is a common item in their lives.

Sewer Lice

They're alive! Or are they? "Sewer Lice" is a fun (if sneaky) way to teach about the behavior of gases in a liquid. It is an attention-getting activity using common household materials that will definitely stimulate interest, generate discussion, and stretch everyone's thinking.

Standards

Science (NRC 1996))   Inquiry, Physical Science, Nature of Science
Math (NCTM 2000)   Reasoning and Proof
Language (NCTE/IRA 1996) Research, Pose Problems, Gather and Evaluate Data, and Communicate Findings; Use Language for Exchange of Information

Integration

Science: Science observations; study of gases and dissolved gases
Math: Data analysis; mathematical terminology
Language: Written records/descriptive writing

SCILINKS.
THE WORLD'S A CLICK AWAY

Topic: Gases
Go to: www.SciLinks.org
Code: DGS65

Objectives

■ Students will make observations.

■ Students will problem solve and explain the discrepant event.

Key Words

buoyancy—the upward force on an object floating in a liquid or gas. Buoyancy allows a boat to float on water.

carbon dioxide—a gas that has no color or odor and is produced whenever anything containing carbon, such as wood or gasoline, is burned. It is breathed out of the lungs of animals and taken in by plants for use in photosynthesis. Carbon dioxide contains two atoms of oxygen for every atom of carbon; its chemical formula is CO_2.

density—a measure of how much matter there is in a certain amount of space. Salt water has a greater density than fresh water.

displacement—the amount of fluid that is pushed out of the way when an object floats in the fluid. The amount of water that is pushed out of the way by a person floating in a pool is the displacement. The weight of that water is also known as the displacement.

Doing Good Science in Middle School   **65**

inference—an assumption arrived at based on examining the evidence that seems to support it

observation—the act of watching something closely and recording how it behaves or changes under certain conditions

skepticism—a doubtful attitude

Topic: Buoyancy
Go to: *www.SciLinks.org*
Code: DGS66

Focus Questions

■ "How could you identify and describe the object I am holding up?"

■ "How can you describe what's happening inside the container on the table?"

Background

In this activity, the wrinkles in the surface of the raisins catch the carbon dioxide (CO_2) bubbles released from the soda. When enough bubbles collect on a raisin, it rises ("swims") to the surface where the bubbles begin to "pop" and the raisin drops or appears to swim back to the bottom of the container.

Tips

■ Use only ice-cold, freshly opened, carbonated cans of soda. (Bubbles are vital to this demonstration, and cans of soda contain more carbonation than plastic bottles.) Mountain Dew works well because it has some color and can be explained as "a nutrient solution to keep the sewer lice alive during shipping and lab time." Other sodas can be used, but the raisins are too easy to identify as raisins when you use a colorless soda such as Sprite, and you can't see the raisins enough to know what they are if you use a cola.

■ Soak the raisins in water for an hour before you need them, or use fresh, moist raisins.

Preparation and Management

■ *Prep time:* 15 min. to prepare sewer lice culture. (Mountain Dew looks like a nutrient solution, and the raisins appear to be small, beetle-like swimming creatures that we call "sewer lice.")

■ *Teaching time:* 45 min. plus 5 min. to pull the lice out of the soda at the end of class and wash up. (*Note:* If you will be using this activity in more than one class, you will want to use an unopened can of ice-cold soda and fresh raisins for each class.)

■ *Materials:*

For teacher demonstration:

■ 1000 mL graduated cylinder or other narrow cylinder (Save a special glass cylinder for this demonstration—clear and clean—and see safety note below.)

■ 1 can Mountain Dew soft drink (ice-cold soda works best)

■ 15–20 raisins

For classroom exploration lab (for each lab group of 3–4 students):

- 1 clear 6–8 oz. plastic cup
- 1 can (12 oz.) Mountain Dew or other soft drink (ice cold)
- snack box of raisins (or equivalent amount)

Procedures

Before you bring out the culture of sewer lice, start with a story—perhaps something like the following:

"One of the biologists from [name a local university or college] *recently discovered a new life form at the mouth of the* [local] *River. The zoologists at the university are still working on the DNA analysis to determine if it is a new species or a mutation of one of the aquatic insects found further upstream. It supposedly was in the newspaper but, truthfully, I never saw the story."*

(If you do this demonstration year after year, you may change the story to explain that you just received the sewer lice you had ordered for their next lab. This works well if you can use a box from a supply company and pour a jar—which you take out of the box—of sewer lice culture into a 1000 mL graduated cylinder. Pour slowly down the side of the cylinder to "minimize injury to the insects.")

Keep the cylinder on a front table at first. Explain:

"I have been able to procure these few for the sake of scientific study with students. Let's take out our lab notebooks and record some initial observations. Some questions you might answer are (1) What are the insects doing? (2) What do they look like (size, shape, color)? (3) What questions can you write down that might help us investigate and learn about these animals?"

After they have recorded their initial observations, you may give additional information.

"Evidently the insect is a mutation of a known life-form, the common louse. Scientists are studying it and its behavior. Although it is a bit early to tell for sure, one of the outcomes they have determined is that this louse is edible and a terrific source of protein! Can you believe it?!

"Another preliminary finding is that these insects also clean up the water and actually help purify it. This is great news in our efforts to clean up pollution in water systems. This particular sample is still a bit unclear but supposedly totally drinkable."

At this point, if you've successfully convinced your students that these are indeed sewer lice, you can reach into the culture, catch a louse, and put it in your mouth. After a moment of hesitation and concern, take a small sip to wash it down.

The confirmation that you successfully convinced your students that the raisins were sewer lice is based on the level of verbal and nonverbal feedback you receive from the students. At this point, you may have a discussion about observations and inferences. We also suggest that you talk with your students about issues of trust. Make sure they understand that the story you told had a purpose and that, intentionally or not, many of our observations are biased by what we believe to be true. In science, we attempt to remove as much bias as possible, but there is always a point at which we have to decide what we are going to believe.

After your demonstration it is time for the students to make their own cultures. (See the safety note below, and reinforce with students at this point that they are not to eat or drink in the science lab). Groups of students use small, clear plastic cups and add fresh soda and a few raisins. Students discuss the "behavior" of the "sewer lice" in the soft drink. They list their observations, questions, and answers to their questions.

As with "The Incredible, Edible Candle," you will want to commit the class to a solemn oath of scientific secrecy. We have found it helpful to tell students they can tell others about the "really cool sewer lice demonstration in science class." However, their friends will have to wait until they have your class to have the opportunity to investigate the amazing behavior of the sewer lice. We would also tell students that if their friends ask too many questions later in the day, they can send their friends to us for answers about the sewer lice and we would perpetuate the story with these students.

Safety Note: In this activity, students should repeatedly be reminded that they should *never* eat or drink from laboratory equipment and that you've used a cylinder that's used *only* for this demonstration every year.

Discussion

■ How does this work? The gas involved is carbon dioxide, and it is dissolved in the soda. This is accomplished by keeping the soft drink container pressurized. Temperature is also a critical factor. The colder the liquid, the more gas that can be dissolved. This is why trout, which require high levels of oxygen in the water, are found in colder water than bass or carp, which don't need as much oxygen and are found in warmer water.

■ The students will see the bubbles on the raisins, and explain the "swimming" to the surface as the addition of bubbles. They may notice an increase in size of some of the bubbles. This may be the addition of more carbon dioxide as it continues to bubble out of the solution (soft drink/carbon dioxide mixture). Ask the students to then explain the evidence for the "sewer lice swimming back down to the bottom."

■ Ask your students to use words such as *density, carbon dioxide, buoyancy,* and *displacement.* The use of technically correct terms minimizes misconceptions and increases scientific literacy.

■ This is an excellent lab to guide students through the process of thinking about their thinking and then have them discuss (or write about) their thinking. Ask them what they thought when you gave your introduction to the demonstration and why they thought that way. For example, they may have been skeptical at first, but as you explained the details of the story, it became more convincing. (What is it about believable details that makes us more willing to believe someone?) Or the students may have believed you because you are the teacher. (This may be the time to build acceptable behaviors for challenging scientific findings.) Then have them compare their thoughts before and after they performed the demonstration/activity themselves—how did it alter their thinking? These are the kind of rich discussions that can develop from thinking about discrepant events.

Extensions (application and inquiry opportunities)

■ Have students repeat the demonstration using a peeled and unpeeled grape; hard, dry raisins; fresh, juicy raisins; soaked raisins—even paper clips. However, we always require students to get approval of their plan before carrying it out. We also require a written proposal (one that we evaluate based partially on whether it is developmentally appropriate to the student). Remind students that all activities are to be carried out in the classroom under teacher supervision.

■ Students who believe they know how to explain this phenomenon can present their ideas to the rest of the class.

Assessment

■ Have students record their observations instead of using their usual experimental science lab report (Appendix B).

■ Have students create posters or diagrams of what happened and why.

Cartesian Diver

We all love brainteasers. Middle schoolers enjoy playing with Cartesian divers, but then we ask them to figure out *why* the divers move. Is it magic? This activity provides students with the opportunity to move through the 5Es—engage, explore, explain, elaborate, and evaluate—to solve the puzzle (Bybee 1997; see pp. 18–19 for a discussion of the model). Students continue to develop their observational skills and practice asking questions and seeking answers—the foundation of scientific inquiry. Additionally, the Cartesian diver lab introduces students to system analysis. (Not to mention, it's fun!)

Standards

Science (NRC 1996)	Inquiry, Physical Science, Nature of Science
Math (NCTM 2000)	Data Analysis and Probability
Language (NCTE/IRA 1996)	Research, Pose Problems, Gather and Evaluate Data, and Communicate Findings

SciLINKS.
THE WORLD'S A CLICK AWAY

Topic: Cartesian Diver
Go to: *www.SciLinks.org*
Code: DGS70

Integration

Science:	Science observations, study of gases, density, and buoyancy
Math:	Data analysis, mathematical terminology, proportionality
Language:	Written records

Objectives

■ Students will make observations.

■ Students will problem solve and explain the relationships between pressure, volume, density, and buoyancy.

Key Words

buoyancy—The upward force on an object floating in a liquid or gas. Buoyancy allows a boat to float on water.

density—a measure of how much matter there is in a certain amount of space. Salt water has a greater density than fresh water.

observation—the act of watching something closely and recording how it behaves or changes under certain conditions

volume—the amount of space that is filled by an object

Focus Questions

- "What is the relationship between pressure and volume of a gas?"

- "How are buoyancy and density affected by the volume of gas trapped in a glass bubble (test tube)?"

- "What happens to the gas molecules when the pressure increases?"

Background

In this activity, the air bubble inside the inverted test tube is compressed (volume decreased) when you squeeze the outside of the capped bottle. When the volume of the bubble decreases, the density of the test-tube system (i.e., the inverted test tube and air bubble) increases because the mass remains the same and the volume of the system decreases. Since the density increases, the test-tube–air-bubble system sinks—it loses its buoyancy. It is important to realize the gas molecules are forced closer together as a result of the increased pressure. You may also ask the students why the water doesn't compress like the gas.

Tips

For each group, prepare one empty 2 L colorless soda pop bottle ahead of time. To remove the labels, we recommend the following technique:

1. Using scissors, cut off the label on the side opposite from the side where the label is glued to the bottle.

2. Fill the bottle with water (to a level above the remaining portion of the label) and screw the cap on snugly.

3. With a blow-dryer, start on low heat (increase heat if needed) and move the blow-dryer along the glued section to evenly warm the area.

4. After 30–60 sec., the glue should be soft enough that the label will peel off easily.

Assembling the Cartesian divers:

1. Fill the 2 L bottle with water to the point where surface tension is evident above the opening—or where the water bulges over the opening of the bottleneck. Put enough water into the 13 mm x 100 mm test tube to fill approximately $1/3$ to $1/2$ of the test tube.

2. Place your finger over test-tube opening and invert the tube.

3. Place the inverted test tube over the 2 L bottle, and slip your finger off the tube as you submerge the test-tube opening under the water in the bottle. (This is a little tricky and requires some practice.) You want the inverted test tube to float, and when you cap the 2 L bottle, some water should overflow so that you end up with a 2 L bottle of water where the only bubble is inside the inverted test tube. (You may need to play with the amount of water in the test tube to get a diver that will sink and float fairly easily.)

Preparation and Management

- *Prep time:* 5 min. per Cartesian diver setup.

- *Teaching time* 35–45 min. plus 5 min. to collect Cartesian diver setups at the end of class.

- *Materials:*

  For teacher demonstration:

  - 2 L colorless soda pop bottle with cap

  - 13 mm x 100 mm test tube

  For classroom exploration lab (for each lab group of 2–3 students):

  - 2 L colorless soda pop bottle with cap

  - 13 mm x 100 mm test tube

- *Procedures:* Before we present the Cartesian diver setups for the students to investigate, we like to start this inquiry with a little discrepancy (aka "magic").

  As you place the Cartesian diver setup on a front demonstration table, ask the students to write down a list of observations (what they actually *see* happening, not inferences or interpretations) in their lab notebooks.

  After they have written a few statements, have each student turn to a neighbor and share their observations. Ask each two-member team to select their two best observations. Then, have the teams share those two observations with two other teams. This group of six students then chooses its first and second choices for best observations. As students then share these, you write them on the board. Once the observations are all listed, ask the students to verify that each statement is an observation, and not an inference. All of this is done with the students in their seats (no one is allowed to come up to closely inspect the Cartesian diver setup). Students copy the class list of observations into their lab notebooks.

  Now demonstrate how the test tube will follow one's finger down and up. (If you place your hand lightly on top of the cap you can exert some force downward and cause the test tube to rise and fall with your finger moving up and down along the side of the bottle, or you can carefully apply pressure by squeezing the bottle.) Have the students write two to three observations. (This is where many students will make inferences, which can help you illustrate how experimenter bias enters into scientific inquiry.)

  Eventually, a student will ask to investigate. This is when we bring out the classroom set of Cartesian divers. Advise students to look closely at the setup and explore the cause-and-effect relationships of the Cartesian divers.

  As the groups of two to three students explore, ask them to develop hypotheses and design possible experiments. Have them make notes of the hypotheses, procedures, results, and conclusions for each test they try. Circulate among the groups and ask them the focus questions on pages 95–96 to guide their inquiry. As the class period draws to a close (or after a second day, depending on the engagement level of the class), ask groups to share their findings with one another and develop a series of statements that address the focus questions. Listing the statements on the board or overhead provides a view of the level of understanding across the class.

Discussion

Use the focus questions for guidance as you move among your scientific research teams.

Extensions (application and inquiry opportunities)

■ Have students swap Cartesian divers, and list the variables that are not consistent among the setups. (Because the students constructed the setups, there will likely be some variation among them—for example, varying amounts of water.) Then ask them if the variables affected the results, and how. Students who believe they know how to explain this phenomenon can present their ideas to the rest of the class.

■ The concepts of density and buoyancy are the same for hot air ballooning; however, temperature is used to change the density of air inside the balloon. Bridging these concepts will enhance the depth of your students' understanding. Tissue paper hot air balloons work well as a follow-up activity to explore the relationship between temperature, volume, pressure, density, and buoyancy. You might even arrange for a hot air balloonist to launch from your school site. Be sure to ask you principal, and follow all school and district policy considerations. It's well worth the effort. The students will be amazed at the size of the balloon and the beauty of balloon flight. You may decide to ask the balloon pilot to visit your class to describe some of the science behind balloon flight, balloon races, and so forth.

Assessment

■ Student lab reports should focus on the following: problem, hypothesis, experimental design, results (data chart, graph, analysis), and conclusion.

■ Students can prepare a poster or diagram of what happened and why.

■ Like Activity #5, this is an excellent lab to guide students through the process of thinking about their thinking, and to have them write about their thinking. Ask them what they predicted as the causes for the movement of the Cartesian diver and why they thought that way—and then compare with what they currently think.

Nut Case

Is every skateboard identical? "No way, dude." Everyone, including students, classifies the things of this world, from skirts to skateboards, but many students may not be aware of the process involved. In this activity, students classify based on observable characteristics. The level of difficulty can be varied in a number of ways according to the choice of nuts.

You may use mixed nuts (in the shell) for initial explorations into classification for younger middle schoolers and sets of the same nuts for older middle schoolers. You may also expand this activity by asking students to develop a classification system for the nuts. (Let your imagination run wild; just don't become a nut case!)

Safety Note: You may want to use beans, bolts, or pasta if you have a student with a nut allergy. Such allergies are still fairly rare, but we recommend that you ascertain whether nuts are safe to use with your students before proceeding… but also be aware of any gluten allergies if you use pasta.

SCiLINKS.
THE WORLD'S A CLICK AWAY

Topic: Classification
Go to: *www.SciLinks.org*
Code: DGS74

Standards

Science (NRC 1996)	Life Science, Inquiry, Nature of Science, Earth and Space
Math (NCTM 2000)	Data Analysis and Probability
Language (NCTE/IRA 1996)	Research, Pose Problems, Gather and Evaluate Data, and Communicate Findings

Integration

Science:	Science observations, classification systems, sorting
Math:	Data analysis, vocabulary
Language:	Written records

Objectives

■ Students will make observations.

■ Students will develop a classification system for a bag of mixed nuts.

Key Word

classification—the grouping of things into an organized system. For example, the classification of living things is based on how they are alike and how they are related to each other.

Focus Questions

■ "How do we organize our world?"

■ "How do we categorize and classify items in our environment?"

■ "How do we select characteristics to use to categorize the things that surround us?"

■ "How do we distinguish between members of similar groups?"

Background

In this activity, we ask students to categorize and classify the nuts they get in a bag of mixed nuts in the shell (or various bags of beans, or a few bags of fifteen-bean soup mix—or other materials, such as bolts or pasta). This is a foundational concept in science that is applied in the classification systems in biology as well as the Earth and space sciences. This activity reveals a lot of information about the cognitive stages of your students. We suggest you ask the students to sort the nuts, and give a creative name to each sorted group. The vast majority of students will sort the nuts by type. Ask students to write out the sorting rules they used. Communication is a critical part of science, and the ability to describe the rules is very important to help others understand the sorting process.

Generally, you want a minimum of choices to make each step in the sorting process. Depending on the level of your students, you may have them share with each other to end up with 20 of the same nuts. Then ask the students to sort these 20 almonds (for example) into groups. (Groups might include, for example, cracked shell, smooth shell, rough shell, length, flatness or thickness of shell, or shade of color.) This is where the activity gets to the heart of the matter. "How do we distinguish between members of similar groups?"

Tips

This activity involves an initial expense (mixed nuts in the shell can be pricey), but the materials have a long shelf life and can be repeated from year to year.

■ Use hard-shell nuts (e.g., pecans, walnuts, filberts) for durability.

■ Students should not attempt to open the shells or eat the nuts. Explain to the students that the nuts are very old, and past the date for eating. The nuts are usually difficult to crack under normal student use. However, prudent classroom safety and management considerations will prevent injuries from shell fragments.

Preparation and Management

■ *Prep time:* 45–60 min. to set up the bags of nuts. Use a minimum of four nut varieties. The more the better. Use 1-gallon zip-closure plastic bags. We use a paper punch to make a few clean holes in the bags. This helps with air circulation for the nuts over the years, and helps the bags of nuts pack into a box for next year.

■ *Teaching time:* 45 min., plus 5 min. to put the nuts back in the bags.

7 activity

- *Materials:*

  For teacher demonstration:

  1 1lb. bag of mixed whole nuts (shells on), available in most grocery stores.

  For additional classroom exploration lab (for each lab group of 2–3 students):

  1 1lb. bag of mixed whole nuts (shells on)

- *Procedures:* Before you begin this activity, engage students in a discussion about how they can tell the difference between different skateboards, soft drinks, or anything else that interests them.

  Once the students are engaged, move into the "Nut Case" activity. You may introduce the activity something like this: "Based on your comments, I think you will develop a clearer understanding of scientific classification if we start with some common objects, nuts. We call it the Nut Case Lab."

Discussion

- Have students diagram the classification scheme for the nuts using Venn diagrams (see Figure 6.5 for an example) to depict the sets and subsets of nuts.

FIGURE 6.5

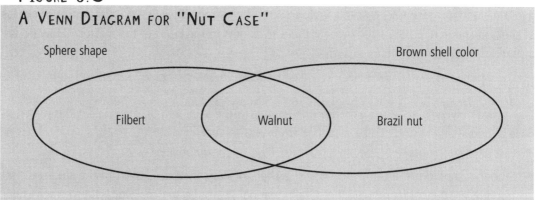

A VENN DIAGRAM FOR "NUT CASE"

Sphere shape | Brown shell color

Filbert | Walnut | Brazil nut

- Ask your students to identify a classification system for something in which they are interested (e.g., celebrities, sports, skateboards, music).

Extensions (application and inquiry opportunities)

- To take the discussion idea further, small groups of students may create expanded classification systems for a number of different groups of objects. For example, they may combine their system for nuts, and add music, candy bars, and so forth—one's edible, one's inedible,....

- Ask groups to explain their expanded classification systems to the rest of the class; have the other class members critique the presentation.

Assessment

■ Have student teams exchange classification systems, and test the system with the other team. How many nuts were correctly classified when students used other teams' classification rules? This can be a springboard for discussing the need for agreed-upon rules for classification, so we can communicate with each other and classify newly discovered organisms. This is where you could introduce a new variety of nut and have students classify it at each step of the classification process.

■ Student lab reports should focus on analysis, classification, and description based on observations (rather than testing a problem).

■ Students can create a poster or diagram of their classification system—perhaps developing a dichotomous key for variety, at a level of complexity appropriate to student ability.

8 *activity*

Wrist Taker

Want to see some puzzled middle schoolers? Ask the following question: "How many *wrists* equal one *neck*?" "Wrist Taker" is an eye-opening activity that provides an introduction to nonstandard units of measurement and why there is a need to have a consistent, common system for measuring and communicating measurements. It also provides opportunity to investigate proportionality and fractions in a unique way. Students (and teachers) are often surprised at the consistent proportionality between parts of the human body, and we've found this lab launches other investigations in the animal kingdom (usually with pet dogs, cats, gerbils, and hamsters or younger siblings as convenient specimens).

Standards

Science (NRC 1996)	Unifying Concepts, Life Science, Inquiry, Nature of Science
Math (NCTM 2000)	Measurement, Algebra (Patterns)
Language (NCTE/IRA 1996)	Research, Pose Problems, Gather and Evaluate Data, and Communicate Findings; Use Language for Exchange of Information

Integration

Science:	Life science (anatomy)
Math:	Measurement, number sense
Language:	Vocabulary, expository writing about procedure and measurements

Objective

■ Students will collect and analyze data and become familiar with nonstandard units of measure.

Key Words

circumference—the line that forms a circle. The length of the circumference of a circle equals 2 times the number pi (π) times the circle's radius (r), or $2\pi r$.

median—(1) the middle number in a sequence of numbers listed from smallest to largest if there is an odd number of numbers. In the sequence 3, 4, 14, 35, 280, the median is 14. (2) the average of the two middle numbers of a sequence of numbers listed from smallest to largest if there is an even number of numbers. In the sequence 4, 8, 10, 56, the median is 9 (the average of 8 and 10).

proportion—the comparative relation between things or magnitudes as to size, quantity, or number

Focus Questions

■ "What is *proportion*?"

■ "Is the circumference of a human wrist proportionate to the size of the neck? Or the size of the head, knee, or ankle?" (Or: "Is there a mathematical relationship between these body parts and the wrist?")

Background

Students will need to have an understanding of the concept of *median* prior to doing this activity. Also, note that

1. The string should be neither stretched nor lax.

2. Most students will have the same proportion of wrists to neck: 1 neck = 4 wrists.

3. After students cut a string equal to each body part listed on the "String 'Em Up" worksheet (p. 82), have them tape the strings to the worksheet. Caution students to line up one end of each string with the line under each subheading at the top of the page.

Preparation and Management

■ *Prep time:* 5 min.

■ *Teaching time:* 45 min.

■ *Materials (per group. We find that groups of four students are a good arrangement.):*

string (approximately 30 in. per student). We discourage waist measurements, which can be embarrassing and which aren't proportional based on bone structure anyway.

scissors

tape

"String 'Em Up" worksheet (p. 82)

■ *Procedures:*

1. Have each group member cut a piece of string to match the circumference of his or her wrist. This will be used to measure other body circumferences.

2. Cut pieces of string to match the circumference of each person's neck. Use wrist strings to measure the length of the neck strings. Record the data on the Body Parts Chart (Figure 6.6).

3. List the neck measurements of each group member in order from "least wrist" to "most wrist."

4. Have students determine—in "wrists"—what the median (or middle) neck size is for each group.

5. Cut a piece of string for the circumference of the body parts listed in the chart. Measure each in "wrists."

FIGURE 6.6
BODY PARTS CHART FOR "WRIST TAKER"

	STUDENT 1	STUDENT 2	STUDENT 3	STUDENT 4	MEDIAN
NECK					
HEAD					
ANKLE					
UPPER ARM					
KNEE					
OTHER					
OTHER					

Discussion

Discuss with students the results of their data. Is there a similarity in their results? Can they see a pattern? Can they make an inference based on the data? Do they need to collect more data? You may want to paraphrase or rephrase some of the following questions:

1. Which strings are nearly the same lengths?

2. Which string when folded in half would be nearly the length of another?

3. Which string when folded in half and in half again (quartered) would be nearly the same as another?

4. Can you find any other body circumference relationships?

You and your students may be astonished to see the consistent proportionality in the human body. This activity is rich with possible discussion topics; be prepared to cultivate interesting digressions and then to bring them back to the lesson objectives.

Extensions (application and inquiry opportunities)

■   You may want your students to continue study of proportion relating to other areas, such as art perspective, or to hypothesize about proportionality in other members of the animal kingdom.

■ Have students construct a ruler with increments of "thumbs" or "hands" and have them measure their desks and classroom. This can be done with a string that has a black mark to indicate each thumb. They may opt to tie a knot for each increment. Display the various "measuring tapes" in the classroom for all students to see.

■ Students can research the origins of measurement and the various methods that were used throughout history and in different cultures.

■ Some middle school students are ready to consider the remarkable findings of scientists studying Fibonacci numbers, Phi, and related examples of proportionality in nature.

Assessment

■ Present individual students (or groups of students) with a predetermined measurement and ask them to determine an equivalent body part.

■ Give students a piece of string. Ask them to find a body part that is half or twice (or whatever) that length. Or, ask them to use the string to measure a specific part and then to explain the mathematical relationship or proportionality they've discovered.

Worksheet:

String 'Em Up!

Name _____

After cutting the strings for the body parts below, tape them on the
line under each subheading.

Neck	Head	Ankle	Upper Arm	Knee	Other	Other

National Science Teachers Association

Oh, Nuts!

Is a peanut just a peanut? This activity, which uses peanuts in their shells, places your students in the role of research scientists who pool their data and explore patterns in nature. They get excited when they discover that patterns exist—in a measly peanut! As with Activity #7, "Nut Case," this activity involves an initial expense but the materials will last for years. "Oh, Nuts!" involves a good amount of math and may need to be adjusted for younger middle schoolers.

Safety Note: As with "Nut Case," due to possible peanut allergy problems that might exist among your students, we recommend that you ascertain whether peanuts are safe to use before conducting "Oh, Nuts!" As a substitute, you may use dried peas or lima, pinto, or navy beans effectively in this activity ("Oh, Beans"?). Remind students that nothing should ever be tasted or eaten in a science room or science lab.

Standards

Science (NRC 1996)	Inquiry, Nature of Science, Life Science, Unifying Concepts
Math (NCTM 2000)	Data Analysis, Probability and Statistics, Measurement, Algebra
Language (NCTE/IRA 1996)	Research, Pose Problems, Gather and Evaluate Data, and Communicate Findings; Use Language for Exchange of Information

Integration

Science:	Developing/standardizing procedures
Math:	Data analysis, using graphs to identify mathematical patterns in nature
Language:	Written records

Objectives

- Students will make measurements.
- Students will analyze data and use graphs to describe patterns in nature (the variation of lengths of peanuts).

■ Students will gain knowledge of procedural consistency.

Key Word

range—the difference between the smallest and largest number in a set of data. If the lowest test score of a group of students is 54 and the highest is 94, the range is 40.

Focus Questions

■ "Are all the peanuts in a bag of peanuts the same length?"

■ "How much variation is there among examples of the same organism?"

■ "How can we make sense when we have a huge amount of data?"

Background

In this activity, students will pool their data as a class. One way to easily collect the class data is to place a chart on the chalk/white board such as the Class Data Chart (Figures 6.7 and 6.8). As students get their measurements completed, they will record their data in their lab notebook data tables as well as the Class Data Chart on the board. (Or save it on a transparency to reuse and refer to later.)

Tips

1. Prepare zip-closure bags with 20 peanuts each. (Or, you may have students select their peanuts from a larger container.) There's a scientific observation to make and share here: "Did students deliberately select peanuts based on certain characteristics?" Discuss with students how this can affect results.

2. We think there are great integration opportunities if you schedule this activity to take place at the same time the students are learning coordinates and graphing in math. The connections are rich and the cross-curricular reinforcement will improve mastery and retention of the concepts.

FIGURE 6.7

CLASS DATA CHART (BLANK) FOR "OH, NUTS!"

RECORD YOUR NUMBER OF PEANUTS WITHIN EACH RANGE							
LENGTH OF PEANUT IN MM	1–10 mm	11–20 mm	21–30 mm	31–40 mm	41–50 mm	51–60 mm	61–70 mm
CLASS TOTALS FOR EACH RANGE							

FIGURE 6.8

CLASS DATA CHART (FILLED IN) FOR "OH, NUTS!"
(WITH CLASS DATA AND TOTALS—FROM 28 STUDENTS X 20 PEANUTS EACH)

RECORD YOUR NUMBER OF PEANUTS WITHIN EACH RANGE						
0	0	1	5	4	0	0
0	0	2	6	1	1	0
0	0	1	6	3	0	0
0	1	1	5	2	1	0
0	0	2	4	3	1	0
ETC.	ETC.	ETC.	ETC.	ETC.	ETC.	ETC.
LENGTH OF PEANUT IN MM 1–10 mm	11–20 mm	21–30 mm	31–40 mm	41–50 mm	51–60 mm	61–70 mm
CLASS TOTALS FOR EACH RANGE 0	36	116	257	110	41	0

Preparation and Management

- *Prep time:* 60 min. to set up the bags of peanuts and rulers.

- *Teaching time:* 45–90 min., plus 5 min. to collect the rulers and bags of peanuts.

- *Materials (per student):*

  20 peanuts in the shell (or other shelled nuts, if there might be peanut allergies among your students)

  1 metric ruler

- *Procedures:*

  1. Present students with various examples of seeds from plants. You may use the nuts you used in Activity #7, "Nut Case," or any other seeds. Then ask the students to group them. Your questions can lead students to discover that all the peanuts, almonds (or any group of the same nut—or lima beans, pinto beans, navy beans, peas) have variations among them.

  2. Then bridge into questions about one seed, the peanut. Present the students with the question "Are all the seeds of the peanut plant alike?" Wait for student input, and then develop a class conversation around the question "How much variation can you identify, and is there a pattern to the variation?" Then expand into a class discussion about the possible differences in lengths of the peanuts.

  3. Pass out the bags and rulers and have students record the length of the peanuts on a data table; then, record their data into the Class Data Chart on the chalk/white board. (If you have not introduced the metric system and measurement, this lab may provide an opportunity to explore standards of measurement, which also ties into Activity #8, "Wrist Taker," and nonstandard measurement.)

4. After the class data are recorded, have students complete their class data chart by drawing up class totals.

5. The data chart is arranged to ease into graphing, and specifically, into histograms. You can go over the basics of constructing histograms and the advantage of using histograms when you have data such as we have in this lab (lots of data that would be very strange to put into a scatter plot, or line graph).

6. After you allow the students to construct graphs using their individual data and class data, engage the students in a conversation about their peanut discoveries as well as their use of graphical analysis to look for patterns in nature.

Discussion

■ When students break a peanut accidentally or otherwise, it's a teachable moment—What do research scientists do with flawed experimental procedures?—because things like broken peanuts happen any time good science is attempted. Scientists record the peanuts that were not used and describe the reasons they were not included in the research.

■ As the class data are recorded, make sure they make sense as you discuss the development of graphs. We usually have students make graphs of their individual data and the class data separately. Once the students have completed both graphs, we discuss the impact of collecting a lot of data compared to just a little data. Sometimes we work this into a conversation about listening to what a friend says you should do in a situation, and listening to what an adult you respect says. Adults usually have a lot more experience—data—upon which to base their decisions.

■ You can have a great class discussion about the value of using math to describe patterns and trends. Graphs usually illustrate data more concisely than a written narrative (paragraph). You may ask the students to describe the class data in narrative form before helping them develop their graphs.

Extensions (application and inquiry opportunities)

■ Students may investigate other samples of living things, such as lengths of oak tree leaves or lengths of any other living thing. Then have them graph the data to see if the pattern we see in the peanuts repeats itself in other sets of measurements. We also have them use their own height—middle schoolers *love* being the data!

■ Students may want to design an investigation to determine if the variation is a function of genetics or environment, or both.

Assessment

■ Science "magazine article": Students write a big, bold headline at the top of their paper, followed by a smaller subheading that tells what their article will be about. They make the article look like a real newspaper or magazine—that is, with columns and perhaps with illustrations or pictures—and they write a story about their discovery. They need to adhere to good journalism—including who, what, when, where, how, and why.

■ Student lab reports: Depending on the level of individual students, they may be able to handle an experimental science lab report with statement of a problem, hypothesis, experimental design, results (data chart, graph, analysis), and conclusion.

■ This is an excellent lab to guide students through the process of thinking about other patterns that might exist in nature. Ask them to recall if they had any ideas about the range of variation of the length of peanuts before carrying out the activity. Having completed the activity, ask what they now know about peanut lengths and how this knowledge would apply to other things in nature (e.g., length of needles on a pine tree, width of leaves from an elm tree, or length of seed pods from a locust tree).

Gobstoppers

This activity is designed to look at properties of objects and materials and changes of properties in matter. The basic process skills of observing, inferring, and communicating are developed along with concepts involved with matter and its changes.

Standards

Science (NRC 1996)	Inquiry, Physical Science, Chemical and Physical Change
Math (NCTM 2000)	Data Analysis and Probability
Language (NCTE/IRA 1996)	Data Recording, Inference, Analysis, Gather and Evaluate Data, and Communicate Findings; Use Language for Exchange of Information

Integration

Science:	Science observations, chemical change, physical change
Math:	Data analysis, vocabulary
Language:	Record data, analyze, infer

Topic: Changes of Properties in Matter
Go to: *www.SciLinks.org*
Code: DGS88

Objectives

- Students will be able to describe matter and its changes.
- Students will be able to infer what happens when matter interacts.
- Students will able to observe and infer patterns of change in matter.
- Students will be able to communicate concepts involved with matter and its change.
- Students will be able to operationally define a mixture.

Key Words

experiment—a test that is done by scientists to find out whether an idea or hypothesis is true

hypothesis—a statement that explains a set of facts and can be tested (by making observations and performing experiments) to determine if it is false or not accurate. Note that a *hypothesis* is not a *prediction*. The following example shows the distinction between the two words: "If a student is doing a science fair investigation to decide what type of paper towel is most absorbent, the student may *predict* that Brand X will hold the most liquid, but this is not a hypothesis. It's a prediction. In this case the hypothesis might be something like 'I think that thicker towels have more spaces to hold liquid.' The student could then go on to say, "Because Brand X is the thickest towel we have, I predict it will hold the most liquid'." (Colburn 2003, p. 95)

variable—a part of a scientific experiment that is allowed to change in order to test a hypothesis

Focus Questions

■ "Does the color of a Gobstopper change?"

■ "How does the color change?"

■ "How quickly does the color change?"

■ "Does the color change more than once?"

Background

Gobstoppers are jawbreakers that change colors and flavors three times. Each dye and flavoring is sealed with several coats of dextrose syrup. The dyes that give this candy its distinct appearance begin as a solution since moisture is needed to develop the color in the dyes. The dye solution is placed in a spray dryer that uses air to drive off the moisture, resulting in dry tablets called *lakes*. These lakes are then attached to aluminum and termed *aluminum lakes*. The aluminum lakes are not water soluble but are water dispersible. It is these dispersions of color that can be seen when the Gobstoppers are placed in water. The aluminum lakes also account for the spectacular patterns observed in the container as the dyes sink in the water due to the fact that they are denser than water.

Preparation and Management

You will need to purchase at least one Gobstopper for each student in your classroom. Gobstoppers are sold nationally by Kmart and Wal-Mart and are usually available at convenience stores such as 7-Eleven as well.

■ *Prep time:* Setup is about 10 min.

■ *Teaching time:* 45 min.

■ *Materials (per group of 4 students):*

    4 different-colored Gobstoppers

    100 mL room-temperature water

    1 clear plastic cup or other clear container

    paper to record observations

    crayons or colored pencils

■ *Procedures:*

1. Explain to the students that they are going to do an experiment with Gobstoppers that will allow them to practice making scientific observations. Have the class think of different ways that scientists might record their observations (e.g., drawings, tables, graphs).

2. Show the students what they will be doing to make sure they understand. Hold up a clear plastic cup. Tell them that theirs will be filled halfway with water. Tell them to gently place four different-colored Gobstoppers into the water, placing them around the edges, as far away from the others as possible, as on a clock face at 12:00, 3:00, 6:00, and 9:00. Stress the importance of not disturbing the container of water.

3. Divide the class into groups of four.

4. Give each group one clear cup and four different-colored Gobstoppers.

5. Tell the students that as soon as they place the Gobstoppers into the water they are to begin making observations. Tell them to write down everything they see and to make at least two drawings as the reaction takes place.

6. Once students begin their experiments, make sure that they are making observations and drawing pictures as the reaction happens.

7. After students have completed their observations (approximately 20 min.) have them individually write at least two questions they had about the reaction. (They tear sheets of notebook paper in half, and write one question on each half.) A sample question: Would this happen if a different liquid were used?

8. When students have finished writing, collect all their questions.

9. Discuss with the students what an experiment is. What is necessary to perform an experiment? Explain that a variable is something in the experiment that you can change to get a different outcome. Ask for examples from the activity (e.g., temperature of water, amount of water, kind of water [distilled, bottled, tap], size of container, number of Gobstoppers, type of liquid).

10. Explore with the students what they think a "testable" question is. Explain that a testable question is one about which you can create an experiment, and change only one variable, in order to find the answer. A testable question cannot be answered by looking up the topic in an encyclopedia! To reinforce the idea of testable questions, explain that you and the class are going to read through the questions they wrote, and decide as a class whether or not each question is testable. It may be helpful to create a chart to which you can tape the questions.

11. During class discussion of testable questions, ask students how they might test their questions—what experiment would they perform? What results would they expect to see?

Discussion

Have groups report their results to the whole class. Consider using or paraphrasing the following questions:

■ What happened to the Gobstopper?

■ What happened to the water?

- How did the Gobstopper and water interact?
- Describe patterns of change in the water.
- Describe patterns of change in the Gobstopper.
- What did you observe to indicate that a Gobstopper is a mixture of different dyes and flavors?

Extensions (application and inquiry opportunities)

- Discuss with students what a hypothesis is, and its function. Explain to students that a hypothesis is an attempt to explain an observation, and that it is *not* a prediction. Provide and discuss examples with the students.

- Explain to students that they are going to design and perform their own experiments using the Gobstoppers. Many students will surely have thought of other experiments they would like to perform while doing the original experiment.

- Ask them to limit their needed materials to common household items to make it easier for you. **Safety Note:** Do not allow the students to use bleach in their experiments. Gobstoppers and bleach react somewhat violently, and it is not suitable for the students to perform such an experiment.

- Do chromatography with M&M's, Skittles, or other sugar-coated colored candies to discover that the different colored coatings are actually three mixtures of pigments. Cut coffee filter strips 2.5 cm x 10 cm. Dip the candy in the water and rub the color off so it makes a line on a coffee filter strip 2 cm from the bottom. Use a pencil to label the top of the strip with the color of the candy. Put water in the bottom of a plastic cup so it is 1 cm deep. Tape the coffee filter strip to a pencil or straw and set the pencil across the top of the plastic cup so the strip hangs into the plastic cup with only the end (below the color line) touching the water. Repeat this procedure for the other different colored candies.

Assessment

The Gobstoppers (or other candy "subject") student lab report can be assessed according to the rubric provided (Figure 6.9).

FIGURE 6.9

RUBRIC FOR "GOBSTOPPERS"

SCORE	CRITERIA
4	**FULL ACCOMPLISHMENT** Student correctly performs the activity. Student is able to observe and record every change in the Gobstoppers. Student is able to observe and record every change in the water.
3	**SUBSTANTIAL ACCOMPLISHMENT** Student correctly performs the activity. Student is able to observe and record most of the changes in the Gobstoppers. Student is able to observe and record most of the changes in the water. Or student is able to observe and record every change in the Gobstoppers and most changes in the water or most changes in the Gobstoppers and every change in the water.
2	**PARTIAL ACCOMPLISHMENT** Student correctly performs the activity but needs help to do it. Student is able to observe and record some of the changes in the Gobstoppers. Student is able to observe changes in only the Gobstoppers or only the water.
1	**LITTLE OR NO PROGRESS TOWARD ACCOMPLISHMENT** Student misunderstands the task or makes little or no effort to perform the activity.

"Gobstoppers" was adapted with permission from an activity created by Dr. Karen Ostlund.

★ ★ ★ ★ ★ ★ ★ ★ ★ ★

Remember that the 10 inquiry-based activities in this chapter are just a few of hundreds, maybe thousands, "out there." Refer to Chapter 8 for resources and leads to locate more activities, and consider how you might transform a traditional lesson you've used in the past to make it more of an active, inquiry-based investigation (as will be illustrated in Chapter 7). Also, take a look at Appendix B, "Sample Lab Report Form."

Have fun! The students will too!

References

Bybee, R. W. 1997. *Achieving scientific literacy: From purposes to practices*. Portsmouth, NH: Heinemann.

Colburn, A. 2003. *The lingo of learning: 88 education terms every science teacher should know*. Arlington, VA: NSTA Press.

National Council of Teachers of English (NCTE) and the International Reading Association (IRA). 1996. *Standards for the English language arts*. Urbana, IL, and Newark, DE: NCTE and IRA.

National Council of Teachers of Mathematics (NCTM). 2000. *Principles and standards for school mathematics*. Reston, VA: NCTM.

National Research Council (NRC). 1996. *National science education standards*. Washington, DC: National Academy Press.

Inquiry Activities in Action

Questioning, Differentiating, and Assessing

Armed with the 10 activities in Chapter 6, or at least the ones that grabbed your interest, let's look at several important considerations for carrying them out. To do that, we will address four topics: (1) focus questions to use to guide inquiry-based activities; (2) ways to "differentiate" the inquiry-based classroom to meet the range of abilities and needs in a typical class; (3) strategies for shifting from traditional teaching (direct-instruction, demonstrations, and lectures) to incorporate a more inquiry-based approach; and (4) assessment of students in an inquiry-based science class.

Questioning strategies in general—and the use of focus questions in particular—are the foundation for inquiry-based instruction. That's where we'll start.

Using Focus Questions

Focus questions may be used in a variety of ways. The teacher can ask them before a demonstration or student lab activity as a form of "anticipatory set" to focus and motivate students. More often, the teacher uses focus questions while he or she is circulating around the classroom during an activity to encourage students to investigate something they have observed, to have students probe an unexpected outcome or discrepant event, or to explain counterintuitive findings. Teachers also use focus questions to ascertain or refresh prior knowledge, to reinvigorate a stalled investigation, to help students discover how they might navigate around misconceptions or errors, and to evoke further questions and inquiries.

The following list is just a short sampling of focus questions and question stems to illustrate this strategy:

- What are you trying to demonstrate or prove?
- What have you determined so far?
- How do you know that?
- Why do you think that happened?
- What didn't work?

- What have you tried? Why? Could we try it another way?

- Is there anything else you could try? What else could you do?

- Is it possible to . . . ?

- Have you considered . . . ?

- What if you . . . ?

- Where could you look . . .?

- Tell me how you . . .?

- What other information do you still need?

- What did you discover when . . .?

- Do we all agree about this?

- What is going on when . . .?

- What do you see? Is that all you see?

- What questions do you have?

- What would happen if . . . ?

- How might that have been different?

- What does that suggest to you?

- Is there someone you could ask?

- Looking back, tell me about . . . .

- What is it exactly that you want to find out?

- What data do we need? How can we collect that data?

Of course, students should be encouraged to generate their own questions, too, and in "full" inquiry (see p. 1) we'd expect them to conduct their own investigations with very little teacher influence. In our experience, inquiry-based instruction in the middle school is most often successful (and practical) when the teacher uses guided inquiry (see p. 1) with focus questions.

The teacher nudges and suggests through questions and brief interviews; attempts at full inquiry can come late in middle school with more experienced youngsters.

Students new to inquiry-based, hands-on science will occasionally hit dead ends; they can also become confused or frustrated when their progress is slow. This is when they need their teacher's help, and focus questions are a key resource for the teacher. Exactly which questions you use will depend on the nature of the students and of their investigation, but it's vital that the teacher is moving around the room, prepared to pose questions at critical moments in the activity. The teacher's goal is to help students gain confidence in their ability to pursue an investigation with increasing independence.

A Differentiated, Inquiry-Based Classroom

Central to a learner-centered teaching approach is the recognition that students acquire, process, and demonstrate mastery of knowledge differently. This is especially true in middle school, where youngsters experience profound developmental changes at rates that vary widely from one student to the next. As we discussed in Chapter 1, it is entirely possible to have a single middle school science class with students functioning at both extremes of the cognitive spectrum—concrete thinking and complex, abstract thinking—and everywhere in-between. For this reason, learner-centered lesson planning calls on teachers to make certain accommodations that will increase the likelihood that good science teaching will take hold (Table 7. 1).

T ABLE 7.1

T RADITIONAL C LASSROOM VS. D IFFERENTIATED C LASSROOM

T RADITIONAL C LASSROOM	D IFFERENTIATED C LASSROOM
Student differences are masked or acted on when problematic.	Student differences are studied as a basis for planning.
Assessment is most common at the end of learning to see "who got it."	Assessment is ongoing and diagnostic to understand how to make instruction more responsive to student needs.
A single definition of excellent prevails.	Excellence is defined largely by student progress from a starting point.
Student interest is infrequently tapped.	Students are frequently guided in making interest-based choices.
Whole-class instruction dominates.	Many instructional approaches are used.
Coverage of text and curriculum guides instruction.	Student interest, readiness, and learning-profile shape instruction.
Single-option assignments are typical.	Multi-option assignments are common.
Single interpretations of ideas and events may be sought.	Multiple perspectives on ideas and events are routinely sought.
A single form of assessment is used.	Students are assessed in multiple ways.

Source: C. A. Tomlinson. 1999. *The differentiated classroom: Responding to the needs of all learners.* Alexandria, VA: Association for Supervision and Curriculum Development. Copyright © 1999 Association for Supervision and Curriculum Development. Reprinted with permission.

Transforming Traditional Lessons

In a traditional science activity, say a pendulum lesson, the teacher and the activity's written directions define how students proceed in a step-by-step manner. How is this different from the inquiry-based learning that we're advocating as good science for middle schoolers? How can you adapt a "cookbook" lesson to include inquiry strategies? Without additional materials, how can you make the necessary subtle shifts to inquiry-based science?

Consider the pendulum lesson in Table 7.2, which shows the differences between traditional and inquiry-based instruction. This example is meant to illustrate how teachers can modify what they're already doing, rather than reinventing entire units of instruction, in order to incorporate more inquiry-based approaches into their daily lessons.

Other steps teachers might take toward an inquiry-based classroom include the following:

■ Provide blank paper instead of worksheets and have students generate

TABLE 7.2

A PENDULUM LESSON: MOVING FROM A TRADITIONAL TO INQUIRY-BASED APPROACH

TRADITIONAL APPROACH: UNDERSTANDING PENDULUMS	INQUIRY-BASED APPROACH: UNDERSTANDING PENDULUMS
Curriculum	Curriculum
■ Presented part to whole; emphasis on basic skills ■ Fixed curriculum ■ Relies heavily on textbooks, worksheets, workbooks *Provide materials to read and worksheet to complete regarding pendulums and motion.*	■ Presented whole to part; emphasis on big concepts and thinking skills ■ Responsive to student questions and interest ■ Relies on collecting primary data and using manipulative materials *Provide each group with possible materials for making a pendulum. Ask students a focus question: "How many times does a pendulum swing?"*
Roles of Students	Roles of Students
■ "Blank slates" onto which information is "etched" by the teacher ■ Work alone *Students follow teacher and worksheet directions exactly: "Cut a piece of string exactly 28 cm long, and tie one end to your pencil and the other end to a washer . . . ."*	■ Thinkers with emerging theories about the world ■ Work in groups *Ask students to construct a pendulum of their own design.*
Role of Teacher	Role of Teacher
■ Generally behaves in a didactic manner; disseminates information to students ■ Seeks the correct answer to validate student learning *Before beginning, instruct students how to hold the pendulum and exactly how to count the swings while the teacher times them.*	■ Generally behaves in an interactive manner; guides but does not direct or mandate during the investigation; allows students to explore, make mistakes, discover on their own; provides direction as needed ■ Sees the students' point of view in order to understand students' present conceptions for use in subsequent lessons *Students swing the pendulum while the teacher times them. Students arrive at different answers for how many "swings" the pendulum made in the time period. Who counted correctly? Why did that happen? Through discussion, students discover "variables" and the need to control them.*
Assessment	Assessment
■ Viewed as separate from teaching; occurs almost entirely through testing *Administer an objective, written test to determine if students understand that the length of string changes the number of swings, while the size of the bob does not.*	■ Interwoven with teaching; occurs through teacher observations of students at work and through student exhibitions and portfolios *Ask students to construct and demonstrate a pendulum that will swing exactly 15 times.*

Adapted from D. Cantrell and P. Barron, eds. 1994. *Integrating environmental education and science.* Newark, OH: Environmental Education Council of Ohio.

their own questions from a chapter reading assignment or teacher demonstration.

- Give students an opportunity to predict, observe, and make sense of data—essentially learn from one another using an "inquiry" skill.

- Give students a variety of materials from which to choose and explore and then encourage students to present questions about and share findings from their varied investigations before the whole class. Meanwhile, audience members can challenge the findings and probe for evidence.

Assessment of Inquiry-Based Science

How do teachers assess learning that takes place during and as a result of a science activity? In this section, we'll share the foundational ideas behind assessment of inquiry-based science in the middle grades and then explore the various forms that assessment can take.

Four premises shape our views on assessing inquiry-based science learning:

1. Due to the wide variations in learning styles, cognitive levels, and maturity among middle school students, we strongly recommend that teachers use preassessment to identify the range and nature of their students' instructional needs and readiness to "do" inquiry-based science at the start of the term and before each unit. At the start of the school year, teachers should determine the following: What do students know about science safety? Do they conceive of science as a process or a collection of discrete knowledge claims? Have they ever engaged in a science inquiry? (We pro-

vide some additional preassessment ideas on p. 100.)

2. Good science assessment is more than a test at the end of the unit or term to find out what students know. It should be used to inform the teacher about what is working and what isn't—*in order to continuously improve teaching and learning.* Therefore, assessment should be diagnostic, varied, and ongoing throughout the school.

3. To help students improve their thinking skills, teachers can shift the responsibility of evaluation and taking action to the students themselves. We want students to learn how to evaluate the quality of their own findings and conclusions and make adjustments as needed based on their self-evaluations. When students are afforded this opportunity, they are thinking like scientists and involved in the highest levels of cognitive activity.

4. Preassessment is more than an event. It is part of an assessment mind-set and is integral to authentic assessment. Teachers who constantly evaluate their students have the information necessary to modify instruction and make curricular decisions that support their students individually. In this way, testing does more than evaluate learning; it also diagnoses the effectiveness of teaching and suggests ways to improve how topics are taught.

Each of these key premises is rooted in the traits of the middle school learner, discussed in Chapter 1, and in the vision of science assessment that emerges from the National Science Education Standards (NSES) (NRC 1996). At its core, assessment of inquiry-based middle

Topic: Assessment Strategies
Go to: *www.SciLinks.org*
Code: DGS99

school science demands that teachers ask students to

- *generate rather than choose a response;*
- *actively accomplish complex and significant tasks; and*
- *solve realistic or authentic problems.* (Layman, Ochoa, and Heikkinen 1996, p. 44)

In addition, teachers should ask students to apply their new learnings to other settings and situations.

The Diagnosis-Prescription Cycle

The effective middle school science teacher begins the year by diagnosing her or his students' needs and then designs activities that cultivate prior experiences and help youngsters at different levels to develop their understanding of science. A diagnostic pretest at the start of the year—ungraded and nonthreatening and presented as a way to see what students know so they can avoid redundant lessons—is a way for teachers to find out what students have retained from previous science instruction. The pretest can include one or more performance-based activities; this will give the teacher a broad view of the remediation, reinforcements, and new teaching that will need to take place in the units ahead. In our pretests, we've used a variety of formats, according to the makeup of each class. Among them are the following:

- Pencil-and-paper test that surveys key concepts and terms students will need in order to master class objectives. (A version of the final unit or term test might be administered as a pretest.)
- Short essay assignment—for example, a "science autobiography" about what they know and want to know about science.

- An assignment that asks for nonverbal representations of science topics and themes, particularly for use with students with language limitations.
- Small-group activity in which students discuss science "facts" provided by the teacher—some factual and some not—and debate the merit of each "fact."
- Round-robin activity with chart paper on the wall listing a different objective on each sheet; lab teams travel from sheet to sheet posting what they know about that objective.

Ideally a pretest series will allow a teacher to observe and evaluate the cognitive, affective, and psychomotor abilities of his or her students. A good pretest is a starting point for unit and lesson planning and should probe what students know about a variety of topics, such as safety, problem solving, key vocabulary, and equipment.

As the school year unfolds, the teacher checks student progress through multiple, ongoing means—for example, journals, portfolios, lab reports, essay and research assignments, student self-assessment, observations, and performance tasks—and adjusts instruction accordingly. This process is called formative assessment; it is used to diagnose weaknesses in student learning and prescribe improvements in teaching, as advocated by Atkin and Coffey (2003). The diagnosis-prescription cycle is unending for effective middle grades teachers. The goal is to appraise the range of skills and aptitudes that are found among middle schoolers. Using a variety of assessment types, like using a variety of instructional methods, is a key to good science in the middle school. As noted in *Classroom Assessment and the National Science Education Standards* (Atkin, Black, and Coffey 2001),

TABLE 7.3

FRAMEWORK OF ASSESSMENT APPROACHES AND METHODS

HOW MIGHT WE ASSESS LEARNING IN THE CLASSROOM?

Selected-Response Format	Constructed-Response Format			
Multiple-choice True/false Matching Enhanced multiple choice	*Brief Constructed Response*	*Performance-Based Assessment*		
	Fill in the blank • Word(s) • Phrase(s) Short answer • Sentence(s) • Paragraphs Label a diagram "Show your work" Visual representation • web • concept map • flow chart • graph/table • illustration	**Product**	**Performance**	**Process-Focused Assessment**

		Essay
		Research paper
		Story/play
		Poem
		Portfolio
		Art exhibit
		Science project
		Model
		Video/ audiotape
		Spreadsheet
		Lab report
			Oral presentation
			Dance/movement
			Lab demonstration
			Athletic skills
			  performance
			Dramatic reading
			Enactment
			Debate
			Musical recital
			Keyboarding
			Teach-a-lesson
				Oral questioning
				Observation
				  ("kid watching")
				Interview
				Conference
				Process
				  description
				"Think aloud"
				Learning log

Source: J. McTighe, and S. Ferrara. 1998. *Assessing learning in the classroom.* Washington, DC: National Education Association. Copyright © 1998 National Education Association. Reprinted with permission.

"because a single piece of work or performance will not capture the complete story of student understanding, assessments should draw from a variety of sources" (p. 57). Consider Table 7.3, which illustrates the range of assessment types available to teachers.

In *Meet Me in the Middle* (2001), Wormeli writes about his experience with the kinds of assessment methods that we have described above. While some assignments might appear "soft" or too simplistic, he says, effective manipulation of content in these methods requires considerable skill on the part of middle grade students. Wormeli describes his success giving a class presentation about how cells die by conducting a funeral for a dead cell, which he deftly entitled, "Death of a Cellsman." Wormeli offers the following alternatives to traditional assessments, all of which can be used at the middle school level in any subject:

Journal or diary entries
Radio plays
Video productions
Debates
Interviews with experts
Annotated catalogs of artifacts
Games and puzzles
Musical compositions
Museum guides
Historical or science fiction stories
Almanacs
News or feature articles
Timelines or murals
Speeches or oral presentations
Advertisements (2001, p. 97)

Assessment Rubrics

Teachers designing assessments of inquiry-based science can use the following process we learned from well-known science educator Bud Alder (1999). (The process is compatible with noted recent works on testing strategies such as Wiggins and McTighe's *Understanding by Design* [1998].) Alder advocates developing rubrics to evaluate inquiry learning using a three-step process. First, the teacher identifies the "enduring learning"—the bottom-line concepts or objectives students must walk away from the activity knowing. Second, the teacher determines what students will actually do to elicit the enduring learning—that is, what the "product" to be assessed will be. Third, the teacher establishes the criteria for evaluation. The same three-step process can be used to design rubrics for a variety of science activities, such as collecting scientific data or creating a science data table. Example of three rubrics developed for the Prince George's County (Maryland) Public Schools appear below. (We received these materials as part of a conference packet in the late 1990s.)

Making Scientific Observations and Drawing Conclusions

Name_____

Date _____

Topic _____

	Assessment	
	Self	Teacher
1. All appropriate senses (*except* taste) were used to make observations.	____	____
2. All appropriate scientific tools and materials used to make observations.	____	____
3. Correct metric measurements were taken when necessary.	____	____

4. Observations were based on what was actually observed and not on personal opinion. ____   ____

5. Collected data are recorded and organized clearly and accurately.   ____   ____

6. Reasonable conclusions were made using observations, collected data, and what was already known.   ____   ____

Scoring Key:

2 points—6 of 6 are checked: Observations and Conclusions Are Excellent.

1 point—4 of 6 are checked: Observations and Conclusions Are Fair.

0 points—3 or fewer are checked: Observations and Conclusions Are Not Acceptable.

Teacher Comments:

Making a Scientific Hypothesis

Name_____

Date _____

Topic _____

	Assessment	
	Self	Teacher
1. My hypothesis is directly related to the question or problem.	____	____
2. My hypothesis is a simple statement based on research and/or what I already know about the question or problem.	____	____
3. My hypothesis states what I believe will happen and why.	____	____
4. My hypothesis is a clear declarative statement.	____	____
5. My hypothesis is written as a complete sentence beginning with a capital letter and ending with a period.	____	____

Scoring Key:

1 point—5 of 5 are checked: Hypothesis Is Excellent.

0 Points—4 or fewer are checked: Hypothesis Is Not Acceptable.

Teacher Comments:

Designing a Scientific Experiment

Name _____

Date _____

Topic _____

	Assessment	
	Self	Teacher

1. The identified question or problem justifies the need for an experiment. ____ ____

2. The design of the experiment tests the hypothesis. ____ ____

3. I included a list of all necessary materials. ____ ____

4. My procedure follows a logical step-by-step order. ____ ____

5. My procedure is written clearly enough that another person could repeat this experiment. ____ ____

6. The procedure shows that repeated trials were completed. ____ ____

7. My experimental design uses a proper control. ____ ____

8. My experiment tests for the effects of only one variable. ____ ____

9. The write-up of the experiment is clear and complete. ____ ____

10. I used complete sentences when writing my question or problem, hypothesis, and procedure. ____ ____

Scoring Key:

3 points: 10 of 10 are checked: Experimental Design Is Excellent.

2 points: 9 of 10 are checked: Experimental Design Is Good.

1 point: 8 of 10 are checked: Experimental Design Is Fair.

0 points: 7 or fewer are checked: Experimental Design Is Not Acceptable.

Teacher Comments:

TABLE 7.4

CHANGING EMPHASES FOR SCIENCE ASSESSMENT

LESS EMPHASIS ON	MORE EMPHASIS ON
Assessing what is easily measured	Assessing what is most highly valued
Assessing discrete knowledge	Assessing rich, well-structured knowledge
Assessing scientific knowledge	Assessing scientific understanding and reasoning
Assessing to learn what students do not know	Assessing to learn what students do understand
End-of-term assignments by teachers and that of others	Students engaged in ongoing assessment of their work
Development of external assessments by measurement experts alone	Teachers involved in the development of external assessments

Source: National Research Council (NRC). 1996. *National science education standards*. Washington, DC: National Academy Press, p. 100.

Metacognition as an Assessment Outcome

The important goal of fostering metacognition—helping students develop an awareness of the quality of their own thinking and course work—is an assessment outcome teachers can achieve in the middle grades by employing the basic inquiry technique of guided questioning. Here are three important prompts:

1. *Where are you trying to go? (Identify and communicate the learning and performance goals.)*

2. *Where are you now? (Assess, or help the student to self-assess, current levels of understanding.)*

3. *How can you get there? (Help the student with strategies and skills to reach the goal.)* (Atkin, Black, and Coffey 2001, p. 14)

Students can revisit these three questions, or variations of them, by keeping a learning log to chart their progress and reflect on their own discoveries.

For many teachers, assessing active learning and metacognition is a departure from their prepa-ration and practice—or a "change in emphasis," as outlined in the National Science Education Standards (NSES) (NRC 1996) (see Table 7.4). The shift in our approach to assessment called for in the NSES is reflected by Bud Alder (1999): "Fifty years ago we assessed what you *knew* to determine what you *could do*. Today we assess what you *can do* to determine what you *know*."

References

Alder, B. 1999. Presentation to Mesa (Arizona) Public Schools science educators. Sept. 27–28.

Atkin, J. M., P. Black, and J. Coffey, eds. 2001. *Classroom assessment and the national science education standards*. Washington, DC: National Academy Press.

Atkin, J. M., and J. E. Coffey, eds. 2003. *Everyday assessment in the science classroom*. Arlington, VA: NSTA Press.

Cantrell, D., and P. Barron, eds. 1994. *Integrating environmental education and science*. Newark, OH: Environmental Education Council of Ohio.

Layman, J., G. Ochoa, and H. Heikkinen. 1996. *Inquiry and learning: Realizing science standards in the classroom.* New York: College Entrance Preparation Board.

McTighe, J., and S. Ferrara. 1998. *Assessing learning in the classroom.* Washington, DC: National Education Association.

National Research Council (NRC). 1996. *National science education standards.* Washington, DC: National Academy Press.

Tomlinson, C. A. 1999. *The differentiated classroom: Responding to the needs of all learners.* Alexan-

dria, VA: Association for Supervision and Curriculum Development.

Wiggins, G., and J. McTighe. 1998. *Understanding by design.* Alexandria, VA: Association for Supervision and Curriculum Development.

Wormeli, R. 2001. *Meet me in the middle: Becoming an accomplished middle-level teacher.* Portland, ME: Stenhouse.

Where Do I Go from Here?

Resources for Good Science in Middle School

At this point, we've looked at the middle school learner, described good science, discussed its classroom implementation, and illustrated it with sample activities. Where can you look to learn more about good science?

About Our Resource Collection

Resources for good science are abundant, and the examples we offer in this section by no means constitute an exhaustive list. These are some of the resources that we've found very useful with a broad audience of science teachers, veterans and novices alike. We believe the very best resource available to teachers seeking to incorporate in-

quiry-based, learner-centered strategies is a mentor in your own school community—someone who has had success with good science instruction in the middle grades—because just like your students, you can learn much faster by seeing and doing it than by reading about it!

On a cautionary note, teachers need to be aware of current copyright restrictions on materials intended for use and distribution at school. Considerations such as brevity of material used, acknowledgment of copyright, and scope of distribution limit what can and can't be used by teachers with students, legally. For updates and details concerning the educational "fair use" regulations, teachers are urged to consult the Library of Congress Copyright Office, which is online at *http://lcweb.loc.gov/copyright*.

The National Standards for Science, Language Arts, and Mathematics

National Science Education Standards (NSES): *www.nap.edu/html/nses/html*

National Standards for the English Language Arts: *www.ncte.org/standards/standards.shtml*

Principles and Standards for School Mathematics: *http://standards.nctm.org*

Publications of the National Science Teachers Association

Crossing Borders in Literacy and Science Instruction: Perspectives on Theory and Practice (2004), edited by E. Wendy Saul. [International Reading Association and NSTA Press; ISBN 0-87207-519-2]

Everyday Assessment in the Science Classroom (2003), edited by J. Myron Atkin and Janet E. Coffey. [NSTA Press; ISBN 0-87355-217-2]

Help! I'm Teaching Middle School Science (2003) by C. Jill Swango and Sally Boles Steward. [NSTA Press; ISBN 0-87355-255-3]

Inquiring Safely: A Guide for Middle School Teachers (2003) by Terry Kwan and Juliana Texley. [NSTA Press; ISBN 0-87355-201-6]

Learning Science and the Science of Learning (2002), edited by Rodger W. Bybee. [NSTA Press; ISBN 0-87355-208-3]

The Lingo of Learning: 88 Education Terms Every Science Teacher Should Know (2003) by Alan Colburn. [NSTA Press; ISBN 0-87355-228-8]

National Science Teachers Association (NSTA) Position Statement: Science Education for Middle Level Students. www.nsta.org/positionstatement& psid=20

NSTA Pathways to the Science Standards, Middle School Edition: Guidelines for Moving the Vision into Practice (1998, 2000), edited by Steven J. Rakow [NSTA Press; ISBN 0-87355-166-4]

Print Resources
Inquiry, Methods, Strategies

Beyond the Science Kit: Inquiry in Action (1996), edited by Wendy Saul and Jeanne Reardon. [Heinemann; ISBN 0-435-07102-5]. Practical, useful advice about providing a rich experience from science kit materials. This book reiterates our assertion that science isn't *in* the kit; it comes *from* the kit. The authors show how to make lessons "real, relevant, and rigorous" using kits.

Bottle Biology (1993, 2003) by Mrill Ingram [Kendall/ Hunt Publishing Co.; ISBN 0-7575-0094-3]. This book includes a wide variety of activities and how-to information for making and using science equipment out of plastic bottles and other recyclable materials. Full of ideas for Earth, life, and physical science applications. See also a related Web site: *www.fastplants.org/bottle_ biology*

Classroom Instruction That Works: Research-Based Strategies for Increasing Student Achievement (2001) by Robert Marzano, Debra Pickering, and Jane Pollock [Association for Supervision and Curriculum Development; ISBN 0-87120-504-1]. An essential resource for educators looking at instructional practices that are grounded in solid research and are shown to have a significant impact on student achievement. Considered a "gold standard" of texts of its sort currently available.

Cooperation in the Classroom (1993) by David W. Johnson, Roger T. Johnson, and Edythe J. Holubec [Interaction Book Co.; ISBN 0-939603-04-7]. This is a foundational work by the authors most closely associated with the contemporary development and spread of cooperative learning. It includes fundamental principles that are still useful almost 15 years after the book was first published.

Cooperative Learning (1994) by Spencer Kagan [Kagan Cooperative; ISBN 1-87097-10-9]. An encyclopedic guide to the theory, methods, and lesson designs of cooperative learning. This book is densely packed with suggestions and useful, readily applicable strategies for increasing student collaboration and group productivity.

The Differentiated Classroom: Responding to the Needs of All Learners (1999) by Carol Ann Tomlinson [Association for Supervision and Curriculum Development; ISBN 0-87120-342-1]. Addresses the fundamental issue of adjusting instruction to student needs and establishing developmentally appropriate instructional objectives. A basic and vital book for effective instruction across grades and subjects.

In Search of Understanding: The Case for Constructivist Classrooms (1999) by Jacqueline Grennon Brooks and Martin Brooks [Prentice Hall; ISBN 0-87120-211-5]. Easy to read and applicable to teachers of all subjects at any grade level, this book is helpful in gaining a fundamental understanding of constructivism and its instructional power.

Inquire Within: Implementing Inquiry-Based Science Standards (2002) by Douglas Llewellyn [Sage Publications; ISBN 0-7619-7745-7]. A reader-friendly overview of inquiry that makes it understandable and applicable in the classroom. An important source for us in writing this book!

Inquiring into Inquiry Learning and Teaching in Science (2002), edited by Jim Minstrell and Emily Van Zee. [American Association for the Advancement of Science; ISBN 0-87168-641-4]. A collection of articles that will be useful for novice and experienced inquiry teachers alike.

Inquiry and Learning: Realizing Science Standards in the Classroom (1996) by John Layman with George Ochoa and Henry Heikkinen [College Entrance Examination Board; ISBN 0-87447-547-3]. This book addresses what standards-based, hands-on/minds-on science instruction looks like in the classroom. The authors encourage questioning, thinking about, and doing science.

Inquiry and the National Science Education Standards: A Guide for Teaching and Learning (2000) by the National Research Council [National Academy Press; ISBN 0-309-06476-7]. An excellent complement to the NSES. Contains vignettes and teaching strategies that teachers will find useful.

Inquiry at the Window: Pursuing the Wonder of Learners (1997) by Phyllis Whitin and David Whitin [Heinemann; ISBN 0-435-07131-9]. Inquiry is everywhere, including the questions that arise from looking out the classroom windows. This book centers on evoking and arousing the curiosity inside young people. Written with an elementary grades perspective, it is relevant to students of all ages pursuing inquiry-learning activities.

Making Sense of Primary Science Investigations (1997) by Anne Goldsworthy and Rosemary Feasey, revised by Stuart Ball [Coordination Group Publications; ISBN 0-86357-282-0]. Don't be put off by the title; this book is anything but "primary." The authors demonstrate how teachers and students have very different perceptions of what science is: Inquiry advocates tend to think we are allowing students to answer their own questions and students see science as a body of knowledge they must learn. This book defines *investigation* and shows teachers how to help students to become investigators and how to fit investigations into the curriculum.

NSTA Pathways to the Science Standards, Middle School Edition: Guidelines for Moving the Vision into Practice (1998, 2000), edited by Steven J. Rakow [National Science Teachers Association; ISBN 0-87355-166-4]. An extremely useful resource for applying the NSES to classroom teaching, this book provides thorough explanations of the concepts, terminology, and objectives set forth in the Standards. A vital resource for teachers new to middle school science.

Nurturing Inquiry: Real Science for the Elementary Classroom (1999) by Charles R. Pearce [Heinemann; ISBN 0-325-00135-9]. Written by a classroom teacher who "walks the talk." Although this book's focus is on the elementary grades, it presents a variety of useful methods that could be used in middle school (e.g., discovery boxes, question boards, and having students write grants to obtain science materials). Pearce is an advocate of student ownership of learning and believes, as we do, that every child is a scientist.

Primary Science: Taking the Plunge (1985) by Wynne Harlen [Heinemann; ISBN 0-435-57350-0]. This book addresses how to get elementary students actively involved with science, including making observations and writing and using questioning strategies. An "oldie but a goodie."

Science for All Children (1997) by the National Science Resources Center (NSRC), in partnership with the National Academy of Sciences and the Smithsonian Institution [National Academy Press; ISBN 0-309-05297-1]. This book presents a rationale for inquiry-centered reform and reviews the building blocks of an effective science program (curriculum, materials support, professional development, assessment, and developing community/administrative support). It also presents insightful case studies of successful inquiry-based programs. It is a must-read for science coordinators and curriculum directors seeking to implement good science at the K–8 level.

Assessment

Classroom Assessment and the National Science Education Standards (2001), edited by J. Myron Atkin, Paul Black, and Janet Coffey. [National Academy Press; ISBN 0-309-06998-X]. Stresses the shift from a traditional view of testing to ongoing formative assessment as set forth in the National Science Education Standards (NSES). A useful companion book for implementing the NSES's vision of assessment.

Understanding by Design (2000) by Grant Wiggins and Jay McTighe [Prentice Hall; ISBN 0-13093-058-X]. A comprehensive and rich study of the connection between learning and meaningful assessment, with particular attention to structuring performance assessments based on their principle of "backward design."

Instructional Strategies for New and Experienced Middle-Level Teachers

The First Days of School: How to Be an Effective Teacher (1998) by Harry K. Wong and Rosemary T. Wong [Harry K. Wong Publications; ISBN 0-9629360-2-2]. Among the most practical and widely used text resources for new and experienced teachers, this book covers every aspect of the crucial first days of class. In middle school especially, the book's emphasis on procedures is a key to success.

Meet Me in the Middle: Becoming an Accomplished Middle-Level Teacher (2001) by Rick Wormeli [Stenhouse; ISBN 1-57110-328-7]. A readable, useful guide to middle school instruction in all subjects, written by a master middle grades teacher and packed with creative suggestions. A must-read for middle grades teachers of all experience levels.

Classroom Management

Conscious Discipline (2001) by Becky Bailey [Association for Supervision and Curriculum Development; ISBN 1-88960-911-0]. Presents a systematic approach to helping students grow through their experience with classroom management procedures. Effective and very appropriate to middle grades learners.

Discipline with Dignity (1999) by Richard Curwin and Allen Mendler [Association for Supervision and Curriculum Development; ISBN 0-87120-357-X]. Presents strategies that are respectful and relevant to teachers working with preadolescents.

Widely read and applied by middle school teachers around the country.

Cognitive Science

Developing Minds: A Resource Book for Teaching Thinking (2001, 3rd ed.), edited by Art Costa. [Association for Supervision and Curriculum Development; ISBN 0-87120-379-0]. The most authoritative and comprehensive review of "brain research" available today. Eighty-five articles cover an extensive range of individual topics that will shape (or re-shape) how teachers perceive the connection between how they teach and how students learn. There are numerous "brain-based" resources available to teachers today, but none are as well-documented scientifically and still relevant to classroom practice as Costa's collection.

Diversity and Poverty

A Framework for Understanding Poverty (2001, rev. ed.) by Ruby K. Payne [aha! Process, Inc.; ISBN 1-92922-914-3]. America's schools reflect the middle class values and training of their teachers and leaders, who are underprepared to serve children living in generational poverty. Payne's book is an insightful companion to her excellent workshop series, a must for school systems dedicated to reaching all children in the quest for student success and in maximizing student achievement. For more about Payne and her work, go to *http://ahaprocess.com*.

Peoples Publishing Group (1-800-822-1080): Our favorite texts from this source are *African and African American Women of Science* [ISBN 1-56256-704-7], *Latino Women of Science* [ISBN 1-56256-705-5], *Ten Great African American Men of Science* [ISBN 1-56256-700-4], and *Multicultural Women of Science* [ISBN 1-56256-702-0]. These four teacher- and student-friendly workbooks include interesting biographies with information and insight often absent from mainstream science textbooks. They also provide struc-

tured hands-on science activities. We feel they are excellent supplemental resources.

Savage Inequalities: Children in America's Schools (1991) by Jonathan Kozol [Perennial; ISBN 0-06-097499-0]. A powerful case study exploring the inequities of public education and the impact of poverty on American children.

Periodicals

National Middle School Association (NMSA) *Middle School Journal* and *Middle Ground* (*www.nmsa.org/services/curriculum.htm*): These journals present crucial topics in middle-level education across content areas and include other topics like management, accountability, leadership, student programs, and more. Available by subscription or as an NMSA member benefit.

National Science Teachers Association (NSTA) member journals *Science Scope* and *Science and Children* (*www.nsta.org/journals*): *Science Scope* is NSTA's journal dedicated to education at the middle and junior high school level; we find the elementary journal, *Science and Children,* to be extremely useful also. NSTA's periodicals feature teacher-tested methods, resources, training, and networking essential to good science instruction.

Web-Based and Multimedia Resources

The Annenberg/Corporation for Public Broadcasting site (*www.learner.org*) has high-quality professional development resources for science and math, including copyright-free reproducibles.

Bottle Biology (*www.fastplants.org/bottle_biology*) is a reliable Web site with some "bottle biology" materials available free as Adobe Acrobat PDF documents.

Carolina Biological (*www.carolina.com*): This site offers useful online teacher resources, including no-cost lesson plans, advice and guidance, teaching strategies, the "Carolina Tips" journal, and more.

Closing the Achievement Gap (*www.teachstream. com*): This video series features noted diversity consultant Glenn Singleton and covers three topics: opening the conversation on race, moving beyond diversity into a greater understanding of race, and taking action to close the achievement gap. Singleton's presentations are among the most powerful we've attended and this video set captures many of his essential themes.

The Exploratorium (*www.exploratorium.com/IFI/ index.html*): The Exploratorium Institute for Inquiry offers outstanding workshops, programs, and online support for inquiry science.

Flinn Scientific (*www.flinnsci.com*): Flinn's comprehensive Web site includes its online catalog with teacher-tested science activities and materials. Free content on the site includes lesson plans and other instructional resources, the "Flinn Fax" quarterly newsletter, and Flinn's extensive, highly regarded science safety support information.

NASA (*www.nasa.gov*): NASA's site offers comprehensive Web-based free resources for teachers. In the past, free professional development from NASA has been arranged for teachers and schools—contact NASA for details.

The Northwest Regional Education Laboratory (NWREL) (*www.nwrel.org/msec/pub.html*) offers many valuable and current free teacher resources, teaching strategies, insights from practitioners, and relevant research on inquiry teaching.

PBS (*www.pbs.org*): The PBS site includes access to high-quality, online interactive simulations, lesson plans, and many free resources for teachers.

Phi Delta Kappa's "Links" (*www.pdkintl.org/links/ linklist.htm*): This site features a variety of resources related to middle-level education organized by PDK, a trusted source of quality educational materials.

The Private Universe (1987) (Videotape produced by Annenberg/CPB: *www.learner.org* or 1-800-532-7637): Why don't students learn science? This provocative video documents and explores the

problem through interviews with teachers, middle school students, and science-illiterate but eloquent Harvard graduates. An eye-opening and ageless resource, this video is still widely used in university courses and in science professional development.

Teacher to Teacher with Mr. Wizard, video series (ECA Educational Services, 1-800-537-0008): Mr. Wizard videos were developed to provide models of good science teaching in action. The programs are intended for use by individual teachers, for staff development, and for preservice at the university level. Emphasis is on K–6 instruction, but all strategies and approaches are transferable to grades 7–12 as well.

TERC Science and Math Learning (*www.terc.edu*): TERC is a nonprofit educational research and development organization dedicated to improving math and science learning and teaching. This site includes high-quality curriculum, professional development and technology information, and software for teachers.

Associations and Workshops

American Association for the Advancement of Science (*www.aaas.org*): AAAS is dedicated to the advancement of science, enhancing cooperative science projects, improving understanding of and funding for science, and contributing to science education. The association's Project 2061 produced *Benchmarks for Science Literacy* (1993), the nation's first comprehensive blueprint for sweeping science education reform, still widely used in school systems today. AAAS publishes the well-known professional journal *Science*.

Association of Science Materials Centers (ASMC) (*www.ces.clemson.edu/aophub*): A LASER (see next entry) affiliate, ASMC, in cooperation with several corporate sponsors, provides the Next Steps Institute (NSI), which focuses on implementation of the five areas of science reform: curriculum, assessment, developing community and administrative support, professional development, and materials support. The NSI is an excellent follow-up training for schools that attended LASER and are dealing with challenges in implementing their science strategic plans.

Leadership and Assistance for Science Education Reform (LASER) (*www.si.edu/nsrc* click on "More about the LASER Center"): The National Science Resources Center, the Smithsonian, the National Academies, and several corporate sponsors offer this intensive one-week institute at a number of regional locations and annually in Washington, D.C. LASER trains teams of educators, community members, and business partners in the five areas of science reform: curriculum, assessment, developing community and administrative support, professional development, and materials support.

National Middle Level Science Teachers Association (*www.nmlsta.org*): An NSTA affiliate serving middle-level science educators with resources and training geared to middle grades teaching and learning.

National Middle School Association (*www.nmsa.org*): NMSA is a central resource for professional development, publications, research, and networking for middle grades teachers in all content areas.

National Science Education Leadership Association (*www.nsela.org*): NSELA provides information to its members on a wide variety of topics (student learning, safety, curriculum, technology, professional development, assessment, inquiry, and science education reform).

National Science Resources Center (*www.nsrconline.org*): NSRC is operated by the National Academy of Sciences, National Academy of Engineering, the Institute of Medicine, and the Smithsonian Institution. NSRC disseminates information about exemplary science teaching resources, develops and disseminates curriculum

materials, and provides training (especially leadership and technical assistance) to promote hands-on science.

National Science Teachers Association (*www.nsta.org*): NSTA is a professional organization for K–16 science educators offering high-quality benefits and resources to members and nonmembers.

Vendors

Activities Integrating Mathematics and Science (AIMS): *www.aimsedu.org*

Biological Sciences Curriculum Study (BSCS): *www.bscs.org*

Center for Science Education, Educational Development Center: *www2.edc.org/cse*

Delta Education: *www.delta-education.com* or 1-800-258-1302.

Foundational Approaches in Science Teaching (FAST), The Curriculum Research and Development Group: *www.hawaii.edu/crdg/FAST.pdf* or 1-800-799-8111.

Full Option Science Systems (FOSS): *www.delta-education.com* or 1-800-258-1302.

Great Explorations in Math and Science (GEMS): *www.lhsgems.org* or *www.carolina.com/GEMS*.

It's About Time, Herff Jones Education Division (includes the *Investigating Earth Systems* video): *www.its-about-time.com*

Kendall/Hunt Publishing Co.:*www.kendallhunt.com* Click on "Middle School"

Science and Technology Concepts for Middle School (STC/MS): *www.nsrconline.org*

Science Education for Public Understanding Project (SEPUP): *www.lhs.berkeley.edu/sepup*

Please note that this section is not a comprehensive listing, nor is it intended as an endorsement of one vendor over another because we have not personally used kits or materials from all of these sources. Readers with questions are encouraged to contact the Association of Science Materials Centers (ASMC) (see "Associations and Workshops," above). The ASMC network includes member schools with individuals who can speak with authority about the specifics and quality of all of these resources (and more).

The Last Word

It's hard to finish writing—or reading—a book like this and feel as if it's really done. That said, we must say that this book is really intended as a beginning. As the preceding resource collection suggests, there are countless opportunities for teachers to extend their skills in good science teaching.

We hope the book serves you as a reference and that you pursue some of the other opportunities outlined in this chapter, particularly the workshops provided by the Exploratorium, NSTA, NSRC, and ASMC. The wealth of learning achieved through interactions with exceptional science educators at such workshops—not to mention the connections made with science educators around the country—has been the most important catalyst in our growth as science educators and learners. That is, next to doing science!

And finally, for those of you just starting to experiment with inquiry-based science activities in the middle grades, remember that the transition is most effective when it comprises subtle shifts rather than a revolution. There will be mistakes along the way, but as Thomas Watson, IBM's founder, was fond of saying: "If you want to increase your success rate, double your failure rate."

Good luck and have fun!

Appendixes

APPENDIX A
Glossary of Good Science Terms

authentic assessment. The use of assessment strategies (beyond paper-and-pencil tests) that involve demonstration of learning—for example, projects, portfolios, and presentations. Authentic assessment is an integral component of good science and middle school instruction.

constructivism. A learning theory centered on the notion that instruction is most effective and lasting when students are actively involved in "constructing" meaning, rather than passively "absorbing" it.

cooperative learning. As with "real" science, good science instruction is most often collaborative; collaborative student groups are most effective when they operate with clear procedures, well-defined expectations for the outcome of their work, and active monitoring by the teacher.

differentiation. Good science instruction is "differentiated" according to the needs and ability levels of students, which can vary significantly among students in a single middle school classroom due to the wide range of cognitive, social, physical, and sexual development of adolescents.

discrepant event. An event that is contrary to what is expected to happen; an anomaly; often the catalyst for an inquiry-based investigation, raising the question "Why did this occur?" and leading to more questions.

extensions. Teachers of good science are encouraged to challenge their students to apply and extend their inquiry investigations, making connections to their everyday lives and pursuing questions of their own in further teacher-approved investigations. Inquiry begets more inquiry.

5E method. Engage/explore/explain/elaborate/ evaluate; a framework for structuring inquiry science lessons.

focus questions. Teachers use focus questions to initiate and guide the progress of inquiry investigations, with the teacher serving as a coinvestigator.

formative assessment. Ongoing assessment, which includes authentic assessment strategies in addition to pencil-and-paper tests and which is used to help teachers gauge student understanding as well as the effectiveness of lessons and activities. Formative assessment must be used to improve instruction for it to be meaningful.

inquiry. An approach to scientific investigation that begins with a question, problem, or observation and is followed by hypotheses testing and the reporting of findings. Often, one inquiry leads to others as learning leads to new questions. Teachers can assume more or less control of an inquiry activity according to difficulty, safety, time, or other concerns.

integration. Good science actively incorporates instruction and practice in the essential skills of mathematics and the language arts. When taught effectively, science is the "crossroads of curriculum."

kits. Also called "units," kits contain materials necessary for students to conduct scientific investigations and are available prepackaged from a number of vendors (see Chapter 8).

metacognition. The awareness of one's learning or thinking processes; teachers help students develop this awareness.

science lab notebooks. Integral to good science, notebooks enable students to record lab observations, practice journaling skills, reflect critically; also, they serve teachers as a means of formative assessment.

scientific method. A cyclic process that may be entered at any point. The authors of this book shy away from defining a "single" scientific method, but the process usually includes observations, hypotheses, problems, experimental designs, data collection and analyses, conclusions, and questions for further investigations.

subtle shifts. A phrase coined by San Francisco's Exploratorium to describe the successful, gradual, low-stress transition from traditional teacher- and text-centered science instruction to active, inquiry-based teaching—teachers are advised to move at a reasonable pace and not attempt (or expect) a revolution in their classrooms.

APPENDIX B

Sample Lab Report Form

Here is a form used by some of our teachers that might be useful to you also:

Name _____   **Date** _____

Problem:

_____

_____

_____

_____

Hypothesis:

_____

_____

_____

_____

_____

Experimental Design:

Results:

Data Chart:

Graph:

Analysis:

Conclusion:

APPENDIX C

NSTA Position Statement: The Nature of Science

Preamble

All those involved with science teaching and learning should have a common, accurate view of the nature of science. Science is characterized by the systematic gathering of information through various forms of direct and indirect observations and the testing of this information by methods including, but not limited to, experimentation. The principal product of science is knowledge in the form of naturalistic concepts and the laws and theories related to those concepts.

Declaration

The National Science Teachers Association endorses the proposition that science, along with its methods, explanations and generalizations, must be the sole focus of instruction in science classes to the exclusion of all non-scientific or pseudoscientific methods, explanations, generalizations and products.

The following premises are important to understanding the nature of science.

- Scientific knowledge is simultaneously reliable and tentative. Having confidence in scientific knowledge is reasonable while realizing that such knowledge may be abandoned or modified in light of new evidence or reconceptualization of prior evidence and knowledge.

- Although no single universal step-by-step scientific method captures the complexity of doing science, a number of shared values and perspectives characterize a scientific approach to understanding nature. Among these are a demand for naturalistic explanations supported by empirical evidence that are, at least in principle, testable against the natural world. Other shared elements include observations, rational argument, inference, skepticism, peer review and replicability of work.

- Creativity is a vital, yet personal, ingredient in the production of scientific knowledge.

- Science, by definition, is limited to naturalistic methods and explanations and, as such, is precluded from using supernatural elements in the production of scientific knowledge.

- A primary goal of science is the formation of *theories* and *laws*, which are terms with very specific meanings.

  1. *Laws* are generalizations or universal relationships related to the way that some aspect of the natural world behaves under certain conditions.

  2. *Theories* are inferred explanations of some aspect of the natural world.

Theories do not become laws even with additional evidence; they explain laws. However, not all scientific laws have accompanying explanatory theories.

3. Well-established laws and theories must:

- be internally consistent and compatible with the best available evidence;

- be successfully tested against a wide range of applicable phenomena and evidence;

- possess appropriately broad and demonstrable effectiveness in further research.

■ Contributions to science can be made and have been made by people the world over.

■ The scientific questions asked, the observations made, and the conclusions in science are to some extent influenced by the existing state of scientific knowledge, the social and cultural context of the researcher and the observer's experiences and expectations.

■ The history of science reveals both evolutionary and revolutionary changes. With new evidence and interpretation, old ideas are replaced or supplemented by newer ones.

■ While science and technology do impact each other, basic scientific research is not directly concerned with practical outcomes, but rather with gaining an understanding of the natural world for its own sake.

—Adopted by the NSTA Board of Directors, July 2000

References

American Association for the Advancement of Science. 1993. *Benchmarks for Science Literacy: Project 2061.* New York: Oxford University Press.

McComas, W., Clough, M., & Almazroa, H. 1998. The role and character of the nature of science in W. F. McComas (Ed.) *The Nature of Science in Science Education: Rationales and Strategies* (pp. 3-39) Boston: Kluwer Academic Publishers.

Moore, J. 1993. *Science as a Way of Knowing: The Foundation of Modern Biology.* Cambridge, MA: Harvard University Press.

National Academy of Sciences. 1998. *Teaching About Evolution and the Nature of Science.* Washington, DC: National Academy Press.

National Association of Biology Teachers. 1987. *Scientific Integrity—A Position Statement.* Reston, VA.

National Science Teachers Association. 1997. *The Teaching of Evolution—A Position Statement of NSTA.* Washington, DC.

APPENDIX D
Science Lab Safety Rules

- Always wash your hands before and after experiments.
- Read all directions for an experiment carefully before beginning. Follow all directions exactly as written or explained by the teacher.
- Never perform unauthorized activities.
- Never mix chemicals or other materials for the fun of it.
- Maintain a clean work area.
- When an experiment is completed, always clean up the work area and return equipment to its proper place.
- Never eat in lab unless authorized to do so.
- Know the location of safety equipment in the lab and how to use it.
- Always wear safety goggles when working with chemicals, burners, or any substance or object that might injure eyes.
- Wear a lab apron when working with chemicals, burners, and other hazardous materials.
- Keep all lids closed when a chemical is not being used.
- Many chemicals and hazardous materials are poisonous. Never touch, taste, or smell any chemical. If instructed to smell the fumes in an experiment, gently wave a hand over the opening of the container and direct the fumes toward the nose.
- Dispose of all chemicals and materials as instructed by the teacher.
- Take care not to spill any materials in lab. If a spill does occur, notify teacher for clean-up directions.
- Be careful when working with acids and bases. Always wear protective gloves when using strong acids or bases. Always pour acid into water when diluting the acid. NEVER pour water into acid.
- Rinse any acids or bases off skin or clothing with water.
- Notify teacher of any acid or base spill.
- Never reach across a flame.
- Keep all materials not used in lab away from flame. Pull back long hair and push up long sleeves if necessary.
- Always point a test tube or bottle being heated away from you and others.
- Never heat liquid in a closed container.
- Always use a clamp, tong, or heat-resistant mitts when handling hot containers.
- Use a wire screen to protect glassware when heating.
- Never heat glassware that is not thoroughly dry.
- Never use broken or chipped glassware. If glassware breaks, notify the teacher.
- Notify the teacher immediately if you are cut in lab.
- All sharp materials and broken glass are to be disposed of in the proper container.
- The gas jets, strikers, and Bunsen burners are to be used properly, as well as any material used to create a flame.

Students are held accountable for their actions in lab. Breaking these rules could have serious consequences, one of which may be loss of lab privileges. This list was originally published in *Help! I'm Teaching Middle School Science* by C. Jill Swango and Sally Boles Steward. 2003. Arlington, VA: NSTA Press.

Bibliography

Alder, B. 1999. Presentation to Mesa (Arizona) Public Schools science educators, Sept. 27–28.

American Association for the Advancement of Science (AAAS). 1989. *Science for all Americans*. New York: Oxford University Press.

American Association for the Advancement of Science (AAAS). 1993. *Benchmarks for science literacy*. New York: Oxford University Press.

Atkin, J. M., P. Black, and J. Coffey, eds. 2001. *Classroom assessment and the national science education standards*. Washington, DC: National Academy Press.

Atkin, J. M., and J. E. Coffey, eds. 2003. *Everyday assessment in the science classroom*. Arlington, VA: NSTA Press.

Atwell, N. 1991. *In the middle: Writing, reading, and learning with adolescents*. Portsmouth, NH: Boynton/Cook.

Blaine, L. 2001. Science is elementary. *CESI Science* 34(2): 17–19.

Bybee, R. W. 1997. *Achieving scientific literacy: From purposes to practices*. Portsmouth, NH: Heinemann.

Bybee, R. W., C. E. Buchwald, S. Crissman, et al. 1990. *Science and technology education for the middle years: Frameworks for curriculum and instruction*. Washington, DC: National Center for Improving Science Education.

Calkins, L. M. 1991. *Living between the lines*. Portsmouth, NH: Heinemann.

Cantrell, D., and P. Barron, eds. 1994. *Integrating environmental education and science*. Newark, OH: Environmental Education Council of Ohio.

Colburn, A. 2003. *The lingo of learning: 88 education terms every science teacher should know*. Arlington, VA: NSTA Press.

Cole, R. W., ed. 1995. *Educating everybody's children*. Alexandria, VA: Association for Supervision and Curriculum Development.

Einstein Project. *Cornerstone study*. 1999. Available online at *www.einsteinproject.org/studies/cornerstone*

George, P. S, and W. M. Alexander. 1993. *The exemplary middle school*, 2nd ed. Philadelphia: Harcourt Brace.

Grennon Brooks, J., and M. Brooks. 1999. *In search of understanding: The case for constructivist classrooms*, rev. ed. Alexandria, VA: Association for Supervision and Curriculum Development.

Jensen, E. 1998. *Teaching with the brain in mind*. Alexandria, VA: Association for Supervision and Curriculum Development.

Johnson, D. W., R. T. Johnson, and E. J. Holubec. 1993. *Cooperation in the classroom*. Edina, MN: Interaction Book Company.

Klayman, R. W., W. B. Slocombe, J. S. Lehman, and B. S. Kaufman. 1986. Amicus curiae brief of 72 Nobel laureates. Available online at *Amicus Curiae of 72 Nobel Laureates*

Klentschy, M., L. Garrison, and O. Amaral. 2001. *Valle Imperial Project in Science (VIPS): Four-year comparison of student achievement data 1995–1999*. Available online at *www.vcss.k12.ca.us/region8/Presentations.html*

Klentschy, M. P., and E. Molina-De La Torre. 2004. Students' science notebooks and the inquiry process.

In *Crossing borders in literacy and science instruction: Perspectives on theory and practice,* ed. E. W. Saul. Wilmington, DE, and Washington, DC: International Reading Association and NSTA Press.

Kohn, A. 1993. *Punished by rewards.* Boston, MA: Houghton Mifflin.

Kohn, A. 1996. *Beyond discipline: From compliance to community.* Alexandria, VA: Association for Supervision and Curriculum Development.

Kozol, J. 1991. *Savage inequalities: Children in America's schools.* New York: HarperCollins.

Layman, J., G. Ochoa, and H. Heikkinen. 1996. *Inquiry and learning: Realizing science standards in the classroom.* New York: College Entrance Preparation Board.

Llewellyn, D. 2002. *Inquire within: Implementing inquiry-based science standards.* Thousand Oaks, CA: Corwin Press.

Lowery, L. 1992. *The biological basis of thinking and learning.* Berkeley, CA: University of California.

Martin, R., C. Sexton, and J. Gerlovich. 2001. *Teaching science for all children*, 3rd ed. Boston: Allyn & Bacon.

Marzano, R. J., D. J. Pickering, and J. E. Pollock. 2001. *Classroom instruction that works: Research-based strategies for increasing student achievement.* Alexandria, VA: Association for Supervision and Curriculum Development..

McTighe, J., and S. Ferrara. 1998. *Assessing learning in the classroom.* Washington, DC: National Education Association.

National Assessment of Educational Progress (NAEP). 2000. Available online at *http://nces.ed.gov/ nationsreportcard*

National Council of Teachers of English (NCTE) and the International Reading Association (IRA). 1996. *Standards for the English language arts.* Urbana, IL, and Newark, DE: NCTE and IRA.

National Council of Teachers of Mathematics (NCTM). 2000. *Principles and standards for school mathematics.* Reston, VA: NCTM.

National Research Council (NRC). 1996. *National science education standards.* Washington, DC: National Academy Press.

National Science Resources Center (NSRC). 1997. *Science for all children.* Washington, DC: National Academy Press.

National Science Teachers Association (NSTA). 1990. An NSTA *Position Statement: Laboratory Science.* Available online at *www.nsta.org/positionstatement &psid=16*

National Science Teachers Association (NSTA) National Convention. 2001. Presentations by the National Science Resources Center (NSRC) and the Einstein Project. St. Louis, MO, March 24.

Payne, R. K. 2001, rev. ed. *A framework for understanding poverty.* Highlands, TX: RFT Publishing.

Saul, W., and J. Reardon. 1996. *Beyond the science kit.* Portsmouth, NH: Heinemann.

Schmidt, W. H., C. C. McKnight, L. S. Cogan, P. M. Jakwerth, and T. R. Houang. 1999. *Facing the consequences: Using TIMSS for a closer look at U.S. mathematics and science education.* Boston: Kluwer.

Schmidt, W., C. McKnight, and S. Raizen. 1997. *A splintered vision: An investigation of U.S. science and mathematics education.* Boston: Kluwer.

Stohr-Hunt, P. 1996. An analysis of frequency of hands-on experience and science achievement. *Journal of Research in Science Teaching* 33:101–109.

Thier, M., and B. Daviss. 2002. *The new science literacy: Using language skills to help students.* Portsmouth, NH: Heinemann.

Third International Mathematics and Science Study (TIMSS). 1997. Available online at *http:// ustimss.msu.edu*

Tomlinson, C. A. 1999. *The differentiated classroom: Responding to the needs of all learners.* Alexandria, VA: Association for Supervision and Curriculum Development.

Vasquez, J., and M. Cowan. 2001. Moving teachers from mechanical to mastery: The next level of science implementation. In *Professional development leadership*, eds. J. Rhoton and P. Bowers, pp. 11–22. Arlington, VA: NSTA Press.

Von Secker, C. 2001. Effects of inquiry-based teacher practices on science excellence and equity. *The Journal of Educational Research* 95(3): 151–159.

Wiggins, G., and J. McTighe. 1998. *Understanding by design*. Alexandria, VA: Association for Supervision and Curriculum Development.

Wise, K. 1996. Strategies for teaching science: What works? *The Clearing House* (July/August): 337–338.

Wolfe, P. 1999. Presentation to Mesa Public Schools, Mesa, AZ, April 16–17.

Wong, H. K., and R. T. Wong. 1998. *The first days of school: How to be an effective teacher*. Mountain View, CA: Harry K. Wong Publications.

Wormeli, R. 2001. *Meet me in the middle: Becoming an accomplished middle-level teacher*. Portland, ME: Stenhouse.

Index

*Page numbers in **boldface** type refer to tables or figures.*